Astrid Sinwel

A New Family of Mixed Finite Elements for Elasticity

Astrid Sinwel

A New Family of Mixed Finite Elements for Elasticity

A Robust Computational Method for Mechanical Problems

Südwestdeutscher Verlag für Hochschulschriften

Impressum/Imprint (nur für Deutschland/ only for Germany)
Bibliografische Information der Deutschen Nationalbibliothek: Die Deutsche Nationalbibliothek verzeichnet diese Publikation in der Deutschen Nationalbibliografie; detaillierte bibliografische Daten sind im Internet über http://dnb.d-nb.de abrufbar.
Alle in diesem Buch genannten Marken und Produktnamen unterliegen warenzeichen-, marken- oder patentrechtlichem Schutz bzw. sind Warenzeichen oder eingetragene Warenzeichen der jeweiligen Inhaber. Die Wiedergabe von Marken, Produktnamen, Gebrauchsnamen, Handelsnamen, Warenbezeichnungen u.s.w. in diesem Werk berechtigt auch ohne besondere Kennzeichnung nicht zu der Annahme, dass solche Namen im Sinne der Warenzeichen- und Markenschutzgesetzgebung als frei zu betrachten wären und daher von jedermann benutzt werden dürften.

Verlag: Südwestdeutscher Verlag für Hochschulschriften Aktiengesellschaft & Co. KG
Dudweiler Landstr. 99, 66123 Saarbrücken, Deutschland
Telefon +49 681 37 20 271-1, Telefax +49 681 37 20 271-0, Email: info@svh-verlag.de
Zugl.: Linz, JKU, Diss., 2009

Herstellung in Deutschland:
Schaltungsdienst Lange o.H.G., Berlin
Books on Demand GmbH, Norderstedt
Reha GmbH, Saarbrücken
Amazon Distribution GmbH, Leipzig
ISBN: 978-3-8381-0704-2

Imprint (only for USA, GB)
Bibliographic information published by the Deutsche Nationalbibliothek: The Deutsche Nationalbibliothek lists this publication in the Deutsche Nationalbibliografie; detailed bibliographic data are available in the Internet at http://dnb.d-nb.de.
Any brand names and product names mentioned in this book are subject to trademark, brand or patent protection and are trademarks or registered trademarks of their respective holders. The use of brand names, product names, common names, trade names, product descriptions etc. even without a particular marking in this works is in no way to be construed to mean that such names may be regarded as unrestricted in respect of trademark and brand protection legislation and could thus be used by anyone.

Publisher:
Südwestdeutscher Verlag für Hochschulschriften Aktiengesellschaft & Co. KG
Dudweiler Landstr. 99, 66123 Saarbrücken, Germany
Phone +49 681 37 20 271-1, Fax +49 681 37 20 271-0, Email: info@svh-verlag.de

Copyright © 2009 by the author and Südwestdeutscher Verlag für Hochschulschriften Aktiengesellschaft & Co. KG and licensors
All rights reserved. Saarbrücken 2009

Printed in the U.S.A.
Printed in the U.K. by (see last page)
ISBN: 978-3-8381-0704-2

For Clemens
and my parents

Preface

In this thesis, a new finite element method to discretize the equations of elasticity is introduced and analyzed thoroughly. As the main result of this work, it is shown that the new method is locking-free with respect to volume and shear locking, i.e. that it is applicable for both nearly incompressible materials and the discretization of the thin structures using flat elements.

To date, several well-known methods for the discretization of the equations of elasticity have been introduced: the primal method using continuous finite element functions for the displacement, but also mixed methods due to Hellinger and Reissner, and variants thereof using weak symmetry. However, each of these methods has its drawbacks, such as locking effects or high computational complexity, which motivates the need for yet another formulation.

The new method lies "in between" the primal method and the mixed Hellinger-Reissner method. The vector-valued displacement function is chosen in the space $H(\text{curl})$, which ensures continuity of the tangential component across interfaces. This choice implies to search for the stresses in the newly introduced space $H(\text{div div})$ consisting of symmetric tensor valued functions, whose divergence again allows for a distributional divergence lying in H^{-1}. It is shown that such tensor fields have their normal-normal component continuous. The resulting mixed formulation is referred to as "Tangential-Displacement-Normal-Normal-Stress (TD-NNS) formulation", and analyzed carefully.

To discretize the TD-NNS formulation, Nédélec elements are used for the displacments, while a new family of symmetric tensor-valued finite elements of arbitrary order is constructed for the stress space. This pair allows for a stable discretization of optimal order of approximation. In computations, hybridization is applied, a technique well known to re-establish positivity of the system matrix for mixed problems. In this setup, stability with respect to volume locking is shown. An additive Schwarz block preconditioner is shown to be an optimal choice, also in case of nearly incompressible materials. Finally, the TD-NNS method is used on thin structures and flat prismatic elements. Stability and error estimates which are independent of the aspect ratio of the elements are derived. All results are confirmed by computational examples.

Writing this thesis would never have been possible without the discussions, inspirations, and helpful hints given by my advisor Joachim Schöberl. Moreover, I owe my thanks to Ragnar Winther for co-refereeing this thesis, and to my colleagues Sabine Zaglmayr and Clemens Pechstein for their time, ideas, and energy. I also thank Clemens and my family for their personal care, support, and encouragement. Last, I want to mention the financial support given by the Austrian Science Fund (FWF) under the START Project Y-192 and by the Austrian Acadamy of Sciences.

Contents

1 Introduction   **7**
   1.1 State of the art 7
   1.2 On this work 11
   1.3 Notations 13

2 Elasticity   **15**
   2.1 Notations 15
   2.2 The equations of elasticity 18
      2.2.1 Kinematics 18
      2.2.2 Balance equations 20
   2.3 Material laws 22
   2.4 Boundary value problems 23

3 Variational Framework   **25**
   3.1 Function spaces 26
      3.1.1 Sobolev spaces on a domain 27
      3.1.2 Sobolev spaces on a manifold 28
      3.1.3 Trace theorems 32
      3.1.4 Green's formulae 35
      3.1.5 Dual space of $H(\mathrm{curl})$ 36
      3.1.6 Inequalities 37
      3.1.7 Regular decompositions of vector fields 38
   3.2 Variational formulations – State of the art 39
      3.2.1 Existence and uniqueness for variational problems 40
      3.2.2 Brezzi's theory of mixed methods 41
      3.2.3 The pure displacement formulation 42
      3.2.4 The Hellinger-Reissner formulation of elasticity 44
      3.2.5 Mixed methods with weak symmetry 45
   3.3 A new weak formulation of elasticity 46
      3.3.1 The space $H(\mathrm{div}\,\mathrm{div})$ 48
      3.3.2 Stability analysis 53

4 Finite Element Methods — 57
- 4.1 Basic ingredients 57
  - 4.1.1 Galerkin approximations of saddle point problems 58
  - 4.1.2 Triangulation 60
  - 4.1.3 Finite element spaces 63
- 4.2 Finite element methods for mixed elasticity 64
  - 4.2.1 Existing methods 65
  - 4.2.2 The TD-NNS method 67
- 4.3 Analysis of the TD-NNS method 74
  - 4.3.1 Stability properties 74
  - 4.3.2 Nodal interpolation and error estimates 81
  - 4.3.3 A Korn-type inequality 87
- 4.4 TD-NNS elements 89
  - 4.4.1 Reference elements and orthogonal polynomials 90
  - 4.4.2 On the triangle 94
  - 4.4.3 On the tetrahedron 97
- 4.5 Numerical results 100

5 A hybrid finite element method — 103
- 5.1 Hybridization of the TD-NNS method 104
  - 5.1.1 Hybridization basics 104
  - 5.1.2 The hybrid TD-NNS method 106
  - 5.1.3 Analysis of the hybrid problem 108
  - 5.1.4 Implementation issues 112
- 5.2 Preconditioning 113
  - 5.2.1 Additive Schwarz block preconditioners 114
  - 5.2.2 A preconditioner for hybridized elasticity 114
  - 5.2.3 Condition number estimates 116
- 5.3 Application to nearly incompressible elasticity 118
  - 5.3.1 Discrete norms 119
  - 5.3.2 Stability 120
  - 5.3.3 A preconditioner for nearly incompressible materials 126
- 5.4 Numerical results 127

6 Anisotropic elements — 131
- 6.1 The TD-NNS method on an anisotropic domain 132
  - 6.1.1 Anisotropic setting 132
  - 6.1.2 Finite element spaces 134
  - 6.1.3 Anisotropic finite elements 135
  - 6.1.4 Stability for tensor-product elements 140

6.2	Interpolation operators	144
	6.2.1 Commuting diagram quasi-interpolation operators	145
	6.2.2 Tensor product interpolation	152
	6.2.3 Anisotropic error estimates	155
6.3	Numerical results	156

Chapter 1

Introduction

1.1 State of the art

Today, the need for computational simulations of mechanical problems arises in many different fields, such as engineering sciences, medicine, or industry. In general, one wants to calculate the deformation of a body under certain loads and the corresponding stress fields. Widely known examples are virtual car crash tests, the simulation of stresses arising in mechanical constructions, but also the optimal design of artificial joints. Often, the problems are not of purely mechanical nature, but coupled, such as the simulation of power transformers, where the arising forces stem from the unknown electromagnetic field.

Mechanical problems can be modeled using partial differential equations (PDEs). The full set of equations for the time-dependent problem consists of certain conservation equations, such as the conservation of mass and momentum, and additional material laws, which have to be chosen for the specific type of solid under consideration. The whole system of PDEs is non-linear, both due to the non-linear nature of the fundamental equations of continuum mechanics, and possibly also to due non-linear material laws. However, in theory as well as in applications, it is necessary to have a good understanding of the underlying linear problem. Linearization of the equations is possible under the assumption of small deformations; one of the simplest material laws is *Hooke's law*, which states a linear dependence between stress and deformation.

Different formulations of the problem of linear elasticity have been derived in the past. Starting from the conservation equations, which are posed in integral sense, it is possible to show equivalence to the *classical formulation*, which consists of partial differential equations, and is well-defined for differentiable stresses and displacements. A very common formulation is the *primal variational formulation*, or *pure displacement formulation*, where the displacements are sought in the Sobolev space H^1 of weakly differentiable functions. The stress tensor is eliminated in this formulation, as it can be expressed in terms of the displacement. One obtains a symmetric, coercive problem, for which existence and uniqueness of a solution can be guaranteed. However, this formulation has its drawbacks: in particular, one may expect stability problems in case of *nearly incompressible materials*. This is due to the fact that the compliance tensor, which links stress and strain, becomes singular in the incompressible limit. The inverse of this tensor acts as a coefficient in the pure displacement formulation, which deteriorates with growing incompressibility of the material. In the incompressible limit, the inverse does not exist, hence the stress tensor cannot be expressed in terms of the displacement. It is not possible to pose the pure displacement formulation in this case.

A possible remedy to stability problems are *mixed variational formulations*, where the stresses are considered as separate unknowns. Here, the compliance tensor itself acts as coefficient. In the framework of mixed problems, it was shown in [ADG84] that then stability estimates do not deteriorate for nearly incompressible materials, the formulation can also be used in the incompressible limit. Due to the singularity of the compliance tensor, compatibility conditions have to be satisfied for certain types of boundary conditions.

The results on existence and uniqueness of solutions in the different formulations as well as the corresponding stability estimates can then be used in more complicated setups. The analysis of the linear case provides the necessary background for both geometrically nonlinear problems, where large deformations are concerned, and non-linear material laws.

Usually, it is not possible to solve the partial differential equations arising from mechanical problems analytically. Therefore, one aims at finding approximate solutions, which solve a discrete system stemming from the infinite-dimensional one. A well known and widely used technique for the discretization of PDEs is the finite element method (FEM). This method is based on the variational formulation of the corresponding PDE. Its main advantage is probably the fact that it is applicable to a huge class of problems, be they linear or non-linear, with varying coefficients, on complex domains with different types of boundary conditions. It has been used and analyzed in different fields, the best-known one the Poisson equation and its variants. Important contributions were made within the context of electromagnetic field problems or fluid mechanics. For an extensive introduction into the finite element method, we refer to [Bra92, BS02].

The general idea of the finite element method is to project the variational formulation posed in some infinite-dimensional space onto a discrete subspace, and solve the corresponding finite-dimensional problem. There are different ideas how to choose these subspaces: probably, the one most commonly used is to take piecewise linear functions on finer and finer meshes. As the mesh size of such triangulations is usually denoted by h, this kind of method is referred to as h-FEM. A different approach is to use piecewise polynomials of increasing order on some prescribed mesh, which results in the p-FEM method (see [BSK81]). In case of analytical solutions, this method results in exponential convergence of the solutions; however, when singularities arise, the h-FEM is better suited to resolve them. The hp-FEM now couples the two approaches, using high order elements, where the solution is smooth, and employing mesh refinement towards singularities. These methods were introduced in [BSK81, BD81]. We refer to the monographs by [Sch98, KS99] for an introduction into hp-FEM, and to [Dem07, DKP$^+$08], where also implementation and automatization of the hp method are treated.

As the number of degrees of freedom in computations is usually large, it is not possible to solve the resulting linear system of equations directly. Instead, iterative solvers such as preconditioned conjugate gradient methods are used. Preconditioning is necessary, as the condition of the system deteriorates with shrinking mesh-size and growing polynomial degree. Symmetry and positive definiteness of the system matrix prove to be of advantage. Different techniques have been developed, and optimal preconditioners for such problems have been designed. When systems stemming from h-FEM discretization are concerned, we mention multilevel and multigrid methods (see [Hac85, BPX90, Bra93, BZ00]), algebraic multigrid methods (see [RS85, RS87]), and domain decomposition methods (see [DW90, TW05]). In context with high-order methods, mostly two-level preconditioners are used. There, one level corresponds to the lowest-order problem, for which either a direct solver, or some optimal preconditioner is applied. For the high-order degrees of freedom, domain decomposition methods induced by the triangulation are common. For details on and examples of such preconditioners, we point the interested reader to [Man90, Sch98, Ain96a, Ain96b, SMPZ08].

There are now several possibilities to define finite element methods for elasticity, which stem from the different variational formulations. Most straightforward is probably to start from the pure displacement formulation and to use piecewise linear functions to approximate the displacement. This approach corresponds to the h-version of the finite element method, and has been analyzed thoroughly, see monographs such as [Bra92, BS02] for details. In general settings, one obtains convergence of the approximate solution to the true one at an optimal rate, as the mesh-size becomes smaller. Preconditioners, such as multigrid or algebraic multigrid methods, have been designed to solve the resulting linear system in optimal complexity, see e.g. [KM87, GOS03, Kra08].

Also for the mixed formulation, finite element methods have been developed, as this formulation shows better stability properties in case of nearly incompressible materials. These methods are often summarized as *Hellinger-Reissner* methods, they go back to [Rei50]. It proves surprisingly hard to construct a pair of finite elements for the displacement vector and the symmetric stress tensor, which satisfies the stability conditions of Brezzi's theory [Bre74, BF91]. In [AW02] a pair of elements in two space dimensions was introduced, the corresponding method in three space dimensions was first characterized in [AC05], a thorough analysis and finite elements were provided in [AAW08]. The construction of these elements in a computational code is not simple, they are of high polynomial order, which implies large costs even for the lowest-order scheme. These problems have been circumvented by yet another mixed formulation, where the symmetry of the stress tensor is enforced only weakly via Lagrangian multipliers. We refer to [Ste86, Ste88, ABD84, AFW07], where such methods were constructed, not all of them conforming. Also the solution of the mixed problem is more involved, we refer to [PW06] for a multigrid method for the conforming mixed problem, which uses the theory developed in [AFW97b, AFW98, AFW00] for the mixed approximation of the potential equation.

Known problems All of the methods mentioned above have their drawbacks, among them locking phenomena such as volume and shear locking. Here, volume locking happens for nearly incompressible materials, while shear locking is usually observed for slim elements. Such elements are natural for the discretization of thin domains, such as plates or shells. In both of these cases, the condition number of the linear system deteriorates, and the computed solutions are of poor quality. We shortly summarize known results and remedies for these problems

- *Nearly incompressible materials.* As one may expect from the analysis in the infinite-dimensional setting, the primal finite element method breaks down, as the material tends to the incompressible limit. Mixed methods are a remedy to this problem; however, we saw that these methods, whether conforming or with weak symmetry, are much more costly than the primal one. For linear elastic, isotropic materials, it is possible to reduce the elasticity problem to a mixed formulation similar to the Stokes problem. Here, not the full stress tensor, but only the scalar-valued pressure is introduced as a separate unknown. This method is shown to be stable also in the incompressible limit. In both cases, the analysis of solution algorithms for the resulting linear system requires additional effort. Apart from [PW06], where multigrid for the full mixed problem is discussed, we refer to [Sch99, Wie00] for suitable preconditioning techniques.

- *Slim domains.* A second difficulty in the discretization of mechanical problems is the aspect ratio of the domain. The standard theory for convergence of the finite element method relies on a shape-regular triangulation of the domain. This means, that the shape of the elements must not deteriorate. When a slim domain, such as a plate or shell, is concerned, the overall mesh size has to be of the order of the thickness of the domain. This implies an enormous amount of elements necessary for discretization. A natural idea is now, to use meshes consisting of flat, prismatic or hexahedral elements, which are aligned with the domain. However, many standard methods fail on such a mesh. This effect is known as *shear locking*. It is closely linked to the constant in Korn's inequality [Nit81, DL76], which deteriorates with the aspect ratio of the domain (here element). In engineering sciences, plate and shell models have been developed to circumvent this problem. There, the system of PDEs on the three-dimensional domain is reduced to a problem on the mid-surface of the structure. Well known are the Reissner-Mindlin [Rei45, Min51] or the Kirchhhoff plate models [TW59]. For shells, the Koiter model [Koi60] is often used. These approaches have been generalized under the notion of hierarchical modeling. There, displacements and/or stresses are assumed to be polynomial in normal direction to the mid-surface. Thereby, one obtains a family of models, as one increases the polynomial order. This may be seen as a *p*-version of a related, three-dimensional finite element method, and was introduced in [VB81]. We refer to [DFY04] for an overview over different discretization methods for plates and shells, with an introduction into hierarchical models and asymptotic expansions. Analysis for different approaches is given in [BL91, AAFM99, DBR01]. A main difficulty related to such methods is the fact, that one additionally obtains unknowns corresponding to rotations or angles of deformation. Thus, coupling these methods to standard ones used in non-degenerated parts of the domain is not straightforward. Also the solution of the underlying system, when direct solvers are not applicable, is less developed than for standard methods. We refer to [PRS90, Pei91, Bre96, AFW97a, SS06], where multigrid methods for plates are analyzed.

1.2 On this work

In this thesis, we present a variational formulation of the elasticity problem, which allows us to construct a finite element method which is stable with respect to volume and shear locking. We propose to use a mixed formulation of the problem of elasticity, where the stresses and displacements are considered as separate unknowns. For the displacement space, we use the space $H(\text{curl})$, which is well known from the analysis and discretization of Maxwell's equations. It can be shown that finite elements for this space must have a continuous tangential component. Widely known as *Nédélec elements* or *edge elements*, they were first constructed in [Néd80]. We refer to the monograph [Mon03] for an extensive introduction into the properties of $H(\text{curl})$ and conforming discretization techniques, and to [Dem07, DKP+08] for the treatment of the hp-version of Maxwell's equations in two and three space dimensions.

The specific choice of $H(\text{curl})$ as displacement space implies a certain space to be used for the stresses. It turns out to be the space of tensor-valued symmetric L^2 functions, whose divergence vector has a divergence, that lies in H^{-1}. We refer to this space as $H(\text{div div})$. Tensor fields in $H(\text{div div})$ have their normal-normal component continuous across interfaces. We thus see that it is necessary to construct tensor-valued symmetric stress finite elements, whose normal-normal component is continuous across element faces. We call the resulting scheme the Tangential-Displacement-Normal-Normal-Stress (TD-NNS) method. We present an analysis of this new problem formulation in both the infinite-dimensional and the discrete setting. We verify the conditions of Brezzi's theory. In the infinite-dimensional setting, this directly implies stability with respect to nearly incompressible materials. To discretize the displacement space $H(\text{curl})$, we use high-order Nédélec finite elements introduced in [Néd80, Néd86]. We explicitly construct shape functions for the stress space of arbitrary polynomial degree greater or equal to one, such that the corresponding finite element method is stable. Using nodal Nédélec interpolation operators for the displacements, and newly defined operators for the stress space, we are able to provide interpolation error estimates with respect to the relevant norms. These estimates then ensure a-priori error bounds, which are of optimal order with respect to mesh refinement.

The resulting discrete linear system is symmetric, but indefinite. We apply a hybridization technique to gain back the positivity, which characterizes the pure displacement formulation of elasticity. Hybridization is a method stemming from domain decomposition. There, the interelement continuity of the flux variable (which are the stresses in our case) is cut, and re-imposed via Lagrangian multipliers living on the interfaces. These multipliers can be seen as approximations to the primal variable. The flux field is then discontinuous, and can be eliminated locally on each element. One ends up with a symmetric positive definite (SPD) system in terms of the Lagrangian multiplier. Such a method was first introduced in [dV65] as an implementation technique for the mixed formulation of Poisson's problem. These techniques were further developed in [AB85], where also an a-priori error bound for the multiplier is given. More recently, hybridization of variable-order Raviart-Thomas and Brezzi-Douglas-Marini elements has been analyzed in [CG04], error estimates can be found in [CG05c].

In the setting of the TD-NNS method, we enforce normal-normal continuity of the stress variable via a Lagrangian multiplier, which then corresponds to the normal displacement. The stress degrees of freedom are local to only one element, and can be eliminated locally. After this elimination, we obtain a SPD system in terms of the displacement quantities (i.e. of the displacement and the Lagrangian multiplier) only. We propose an additive Schwarz preconditioner to use in a preconditioned conjugate gradient method, and prove its spectral equivalence to the SPD system. We show that the hybridized problem is well-suited for the treatment of nearly incompressible materials. In that case, we propose to add a consistent, local stabilization term. Then, the condition of the preconditioned system can be shown to be independent as well of the mesh size, as of the material tending to the incompressible limit. We provide error estimates for the displacement quantities, which are independent of the Poisson ratio of the material, which tends to 1/2 for nearly incompressible materials.

A major task is the application of the TD-NNS method to anisotropic domains, such as plates or shells. We propose to use flat quadrilateral or prismatic elements aligned with the domain. Prismatic or quadrilateral, $H(\text{curl})$ conforming elements are well known, see e.g. [Néd80, Néd86, Mon03, DKP$^+$08]. We explicitely construct finite element bases on for corresponding stress finite elements. We provide stability and approximation estimates, which are independent of the aspect ratio of the elements. This is possible, as we use a broken H^1 norm for the displacements, which contains only the strain tensor, and not the full gradient of the displacement. Also interpolation error estimates are provided with respect to the strain tensor only. This way, we are able to avoid the use of Korn's inequality on the element. Thus, it is possible to use flat elements of high order in the in-plane direction, which allows us to save a considerable number of degrees of freedom compared to standard methods, where a shape-regular mesh of overall mesh size similar to the thickness of the domain has to be used. In comparison to standard plate or shell models, we retain more flexibility regarding the polynomial order in our method, as also coupling to well-shaped parts of the computational domain can easily be done.

The TD-NNS method, as described in this thesis, has been implemented in the finite element open-source software package Netgen/NgSolve

http://www.hpfem.jku.at/.

The proposed stress elements are available of arbitrary order and for all types of elements considered in this work. All computations have been done using this code.

This thesis is organized as follows.

In *Chapter 2*, the fundamental equations of continuum mechanics are presented. The material law of linear elasticity (Hooke's law) is stated, and the notion of (near) incompressibility is introduced. Different kinds of boundary conditions to guarantee unique solvability of the underlying boundary value problem are prescribed.

In *Chapter 3*, first a background concerning Sobolev spaces, their properties and trace maps, is given. Then, variational formulations for the problem of mixed elasticity are deduced. Among them, well-known are the primal pure displacement, the Hellinger-Reissner, and mixed-symmetry schemes. Last, we present the TD-NNS formulation, which is analyzed subsequently.

Chapter 4 deals with finite element methods for elasticity. After a short summary on basic notions concerning FEM for mixed problems, we state existing methods, before proceeding to the development of finite elements for the TD-NNS method. A thorough analysis of stability and a-priori error estimates follows.

Hybridization techniques are considered in *Chapter 5*. There, we deal with preconditioning for the positive definite problem. Subsequently, the method is applied to nearly incompressible materials, where independence of the degree of incompressibility can be shown.

Chapter 6 is devoted to the analysis of the TD-NNS method on anisotropic domains. Tensor product elements of arbitrary order are constructed and analyzed. Error estimates independently of the aspect ratio of the finite elements are given.

1.3 Notations

In this thesis, we use the following symbols and notations. Precise definitions of non-standard items are given at the indicated pages. Let $D \subset \mathbb{R}^d$, $d = 2, 3$ be a domain, $A \subset \partial D$ a part of its boundary, v, w vector fields and σ a tensor field of second order.

$v \cdot w$	inner product of two vectors
$v \otimes w$	tensor product vw^T of two vectors
v_n, v_τ	normal and tangential component of v on surface with normal n; $v_n = v \cdot n$, $v_\tau = v - v_n n$
σ_n	surface vector of a tensor field, $\sigma_n = \sigma n$
$\sigma_{nn}, \sigma_{n\tau}$	normal and tangential component of σ_n; $\sigma_{nn} = n^T \sigma n$, $\sigma_{n\tau} = \sigma_n - \sigma_{nn} n$
$n^\perp$	rotation in two dimensions $(-n_2, n_1)^T$
$\mathbb{R}_{SYM}, \mathbb{R}_{SKW}$	the sets of symmetric and skew-symmetric matrices
$\lvert \cdot \rvert$	absolute value of a scalar, Euclidean norm of a vector, and Frobenius norm of a matrix-valued quantity
$\lambda_{min}(\sigma), \lambda_{max}(\sigma)$	smallest and largest eigenvalue of a matrix
$\lvert \sigma \rvert_s$	spectral norm of a matrix, $\lvert \sigma \rvert_s = \lambda_{max}(\sigma^T \sigma)$
V^*	dual space of V

A^*	adjoint operator of A;
$P^k(D)$	polynomial space on the domain D of order at most k, page 65
$P_0^k(D)$	subspace of $P^k(D)$ satisfying zero boundary conditions on ∂D, page 65
$Q_{x_1,\ldots,x_n}^{k_1,\ldots,k_n}(D)$	polynomial space of mixed order, $Q_{x_1,\ldots,x_n}^{k_1,\ldots,k_n}(D) = P_{x_1}^{k_1} \otimes \cdots \otimes P_{x_n}^{k_n}$, page 136
$\mathcal{C}^k(D), \mathcal{C}^\infty(D)$	space of k-times/infinitely often continuously differentiable functions on D, page 18
$\mathcal{C}_0^k(D), \mathcal{C}_{0,A}^k(D)$	subspaces of $\mathcal{C}^k(D)$ satisfying homogenous boundary conditions for the function and its first $k-1$ derivatives on ∂D or the boundary part $A \subset \partial D$, respectively, page 28
$\mathcal{C}_0^\infty(D), \mathcal{C}_{0,A}^\infty(D)$	subspace of $\mathcal{C}^\infty(D)$ of functions of compact support in the open domain D, or $D \cup A$ respectively, page 28
$H^k(D)$	Sobolev space of order k, page 27
$H_0^k(D), H_{0,A}^k(D)$	subspaces of $H^k(D)$ satisfying homogeneous boundary conditions for the function and its first $k-1$ derivatives on ∂D, A respectively, page 28
$H(\mathrm{curl}; D)$	Space allowing for a weak curl operator, page 27
$H(\mathrm{div}; D)$	Space allowing for a weak divergence operator, page 27
$P^k(\mathcal{D}), H^k(\mathcal{D})$	for a domain decomposition $\mathcal{D}$ are defined piecewise without continuity assumptions, page 65
$V \otimes W$	tensor product of spaces, page 17
$a \simeq b$	similarity of $a, b \in \mathbb{R}$, there exist constants $c_1, c_2 > 0$ such that $c_1 a \leq b \leq c_2 a$, possible (in)dependencies of c_1, c_2 are pointed out

Chapter 2

Elasticity

This chapter is devoted to the basic principles of continuum mechanics. We recall the theory as provided in standard literature on elasticity, such as [MH94, Cia88, FdV79]. Starting from some reference configuration of the body of interest, we first define the notions of deformation, displacement, velocity and acceleration. We further introduce the strain tensor, which describes the change of length and angles in the body. Under the assumption of small deformations, we identify reference and deformed configuration. It is then rectified to replace the non-linear strain tensor by its linearization.

The balance of mass and momentum are the basic equations of kinematics. For the stress tensor, the balance of angular momentum implies its symmetry. In the steady state case, the conservation of mass and momentum reduce to the equilibrium equation, which relates the divergence of the stress tensor with the given volume forces.

We further discuss material laws, which relate stress and strain, and thereby characterize also the relation between stress and displacement. We restrict ourselves to the case of linear elastic materials, which are governed by Hooke's law. It states a linear dependence between the stress and strain tensor. A homogenous, isotropic material is determined by two constants. In the classical theory of mechanics, one mostly uses Young's modulus and Poisson's ratio. We introduce the notion of nearly incompressible and incompressible materials in this case.

In Section 2.1, we introduce some standard vector, matrix and differential operators. In Section 2.2, we provide the governing equations of continuum mechanics, and their linearization. Section 2.3 contains material laws for elastic bodies, while in Section 2.4 the boundary value problem of linear elasticity is stated.

2.1 Notations

In this thesis, we do not differ in our notation between scalar, vectorial and tensor-valued quantities, and we identify matrices and tensors of second order. Let $d \in \mathbb{N}$ be fixed, and let $v \in \mathbb{R}^d$ be a

d-dimensional vector. We refer to its i-th component as v_i, and denote

$$v = (v_1, \ldots, v_d)^T.$$

A vector can also be seen as $d \times 1$-matrix. For general $d_1, d_2 \in \mathbb{N}$, $\gamma \in \mathbb{R}^{d_1 \times d_2}$ is a tensor of dimension $d_1 \times d_2$, with components γ_{ij}. A quadratic tensor $\gamma \in \mathbb{R}^{d \times d}$ is called symmetric, iff it coincides with its transpose, $\gamma = \gamma^T$, and skew-symmetric, iff $\gamma = -\gamma^T$. We refer to the respective real spaces as $\mathbb{R}^{d \times d}_{SYM}$ and $\mathbb{R}^{d \times d}_{SKW}$. If the dimension is clear from the context, we simply write $\mathbb{R}_{SYM}$ and $\mathbb{R}_{SKW}$.

We use the dot for the inner product between vector fields $v, w \in \mathbb{R}^d$, and the column for the matrix inner product between $\gamma, \delta \in \mathbb{R}^{d_1 \times d_2}$

$$v \cdot w := v^T w = \sum_{i=1}^{d} v_i w_i, \qquad \gamma : \delta := \sum_{i=1}^{d_1} \sum_{j=1}^{d_2} \gamma_{ij} \delta_{ij}.$$

In case of $d = 3$, we introduce the cross product

$$v \times w := \begin{pmatrix} v_2 w_3 - w_2 v_3 \\ v_3 w_1 - w_3 v_1 \\ v_1 w_2 - w_1 v_2 \end{pmatrix}.$$

The tensor product $\otimes$ shall denote the outer product between two vectors $v \in \mathbb{R}^{d_1}$, $w \in \mathbb{R}^{d_2}$

$$v \otimes w := v w^T \in \mathbb{R}^{d_1 \times d_2}.$$

We use $|\cdot|$ to denote the absolute value of a scalar, as well as the Euclidean norm of a vector, and the Frobenius norm of a tensor. For a domain or surface D, $|D|$ shall be its volume or area, respectively. The spectral norm $|\cdot|_s$ of a second order tensor is defined by

$$|\gamma|_s := \sqrt{\lambda_{max}(\gamma^T \gamma)},$$

where $\lambda_{max}(\delta)$ denotes the maximal eigenvalue of the matrix δ.

Let now I be the identity matrix of dimension d. For a quadratic tensor $\gamma \in \mathbb{R}^{d \times d}$, we define the trace and deviator by

$$\mathrm{tr}(\gamma) = \gamma : I, \qquad \mathrm{dev}(\gamma) := \gamma - \frac{1}{d} \mathrm{tr}(\gamma) I.$$

Note that there holds $\mathrm{tr}(\mathrm{dev}(\gamma)) = 0$.

Let now $d, k \in \mathbb{N}$ be fixed, and $x = (x_1, \ldots, x_d)^T$. We denote the set of polynomial functions in x up to order k on $\mathbb{R}^d$ by $P_x^k(\mathbb{R}^d)$. For $k_1, \ldots k_d \in \mathbb{N}$, the space $Q_x^{k_1, \ldots, k_d}(\mathbb{R}^d)$ shall be the polynomial

space of mixed order,
$$Q_x^{k_1,\dots,k_d}(\mathbb{R}^d) := P_{x_1}^{k_1}(\mathbb{R}) \otimes \cdots \otimes P_{x_d}^{k_d}(\mathbb{R}).$$
There, the tensor product $\otimes$ between spaces is defined by
$$V \otimes W := \operatorname{span}\{vw \mid v \in V, w \in W\}.$$
If clear from the context, the subscript x can also be omitted. By $P_{SYM}^k(\mathbb{R}^d)$ and $P_{SKW}^k(\mathbb{R}^d)$, we abbreviate the spaces of tensor-valued, $d \times d$-dimensional, symmetric and skew-symmetric polynomials, respectively.

Let now $D \subset \mathbb{R}^d$ be an open domain with smooth boundary. By $P^k(D), Q^{k_1,\dots,k_d}(D), P_{SYM}^k(D)$ and $P_{SKW}^k(D)$ we mean the respective spaces of polynomial functions restricted to D. Next, we consider some smooth function $w : D \to \mathbb{R}$. We define the gradient operator $\nabla := (\frac{\partial}{\partial x_1}, \dots, \frac{\partial}{\partial x_d})^T$
$$\nabla w := \left(\frac{\partial w}{\partial x_1}, \dots, \frac{\partial w}{\partial x_d}\right)^T.$$
For a vector-valued function $v : D \to \mathbb{R}^d$, its gradient is a tensor, where each of its rows is the gradient of the respective component of v,
$$\nabla v := \begin{pmatrix} \nabla v_1^T \\ \vdots \\ \nabla v_d^T \end{pmatrix}.$$
In two dimensions, there exist two curl operators: the vector Curl mapping scalar to vector fields, and the scalar curl operator acting vice versa,
$$\operatorname{Curl} w := \left(-\frac{\partial w}{\partial x_2}, \frac{\partial w}{\partial x_1}\right), \qquad \operatorname{curl} v := \frac{\partial v_1}{\partial x_2} - \frac{\partial v_2}{\partial x_1}.$$
In three space dimensions, there is only one curl operator, which maps vector fields to vector fields
$$\operatorname{curl} v := \operatorname{Curl} v := \nabla \times v.$$
The divergence of a vector-valued function q is defined as
$$\operatorname{div} q := \nabla \cdot q = \sum_{i=1}^d \frac{\partial q_i}{\partial x_i}.$$

For a tensor-valued function $\tau : D \to \mathbb{R}^{d \times d}$, $d = 2, 3$, curl and divergence are defined row-wise

$$\operatorname{curl} \tau := \begin{pmatrix} \operatorname{curl}(\tau_{11}, \ldots, \tau_{1d}) \\ \vdots \\ \operatorname{curl}(\tau_{d1}, \ldots, \tau_{dd}) \end{pmatrix}, \quad \operatorname{div} \tau := \begin{pmatrix} \operatorname{div}(\tau_{11}, \ldots, \tau_{1d}) \\ \vdots \\ \operatorname{div}(\tau_{d1}, \ldots, \tau_{dd}) \end{pmatrix}.$$

They map to $\mathbb{R}^{d \times d(d-1)/2}$ and $\mathbb{R}^d$, respectively. Throughout this work, we denote

- $\mathcal{C}(D)$ the sets of continuous functions on D,
- $\mathcal{C}^k(D)$, $\mathcal{C}^\infty(D)$ the set of k times/infinitely often continuously differentiable functions on D,
- $\mathcal{C}_0^k(D)$ the subset of $C^k(D)$ with vanishing traces of the function and its first $k-1$ derivatives on the boundary ∂D,
- $\mathcal{C}_0^\infty(D)$ the set of functions in $\mathcal{C}^\infty(D)$ with compact support.

Given a non-overlapping domain-decomposition $\mathcal{D} := \{D_1, \ldots, D_S\}$ of D into sub-domains D_i, we define all above spaces also for this decomposition. There, we do not pose any assumptions on the continuity between the different domains, i.e. the spaces are defined piecewise. For example $P^k(\mathcal{D}) = \{v : D \to \mathbb{R} \mid v|_{D_i} \in P^k(D_i) \text{ for } i = 1, \ldots, S\}$.

2.2 The equations of elasticity

In this section, we introduce the basics of continuum mechanics. Let $\Omega \subset \mathbb{R}^d$, $d = 2, 3$ be a domain in space, which is identified with the body of interest at some initial time. We call Ω the *reference configuration*. We provide a framework for the description of the motion of Ω in time and space under given surface and volume forces. We always assume that Ω is open and connected. So far, let the boundary of Ω be smooth, we will consider more general settings later in this thesis.

2.2.1 Kinematics

Kinematics is the mathematical description of the deformation and motion of a piece of material, cf. [MH94]. Let $\Omega \subset \mathbb{R}^d$, $d = 2, 3$ be identified with this piece at time $t_0 = 0$. We are then interested in its movement in the time interval $[0, T]$. We call a smooth mapping $\Phi : \Omega \to \mathbb{R}^d$ a *configuration* or *deformation* of Ω, if it preserves orientations and is invertible, i.e. if

$$\det(\nabla_X \Phi(X)) > 0 \quad \text{for all } X \in \Omega.$$

A point $X \in \Omega$ is called *material point*, while some point $x \in \mathbb{R}^d$ is called *spatial point*. We will denote quantities related to the reference configuration by uppercase letters. For the same quantity on the deformed configuration, the respective lowercase letter is then used, e.g. the material point X matches the spatial point $x = \Phi(X)$.

A time-dependent, continuous family of deformations $(\Phi(\cdot, t))$ is called *motion*. At time t, any $X \in \Omega$ is mapped to some spatial point $x = \Phi(X,t)$. For such a point $x = \Phi(X,t)$ we define the

$$\begin{aligned}
&\textit{displacement} & U(X,t) &:= \Phi(X,t) - X, & u(x,t) &:= U(X,t),\\
&\textit{velocity} & V(X,t) &:= \frac{\partial}{\partial t}\Phi(X,t), & v(x,t) &:= V(X,t),\\
&\textit{acceleration} & A(X,t) &:= \frac{\partial^2}{\partial t^2}\Phi(X,t), & a(x,t) &:= A(X,t),
\end{aligned}$$

with respect to material and spatial coordinates.

We now introduce the notion of strain. Let $\nabla_X \Phi$ be the gradient of Φ with respect to X, which is referred to as the *deformation gradient*. The local deformation of the medium is characterized by the *right Cauchy-Green tensor*

$$C(X,t) := \nabla_X \Phi(X,t)^T \nabla_X \Phi(X,t),$$

or, equivalently, *Green's strain tensor*

$$E(U)(X,t) := \frac{1}{2}(C(X,t) - I) = \frac{1}{2}\Big(\nabla_X U(X,t) + \nabla_X U(X,t)^T + \nabla_X U(X,t)^T \nabla_X U(X,t)\Big).$$

Note that these tensors are nonlinear with respect to deformation or displacement. A detailed analysis of the properties of these tensors can e.g. be found in [MH94]. We state that the strain vanishes, if the motion is a rigid body transformation, i.e. if

$$\Phi(X,t) = Q(t)X + b(t)$$

for some smooth vector field $b : [0,T] \to \mathbb{R}^d$ and a smooth, orthogonal tensor field $Q : [0,t] \to \mathbb{R}^{d \times d}$, $Q(t)^T Q(t) = I$. For such a motion, Q corresponds to the rotation, while b is related to the translation of the body.

So far, we provided basic notions of continuum mechanics for arbitrary motions. In the following, we will restrict ourselves to "small deformations". Thus, it is rectified to identify the material point X with its image x at time t. We can replace material coordinates by spatial ones, setting $X = x$. We will use $u(x,t), v(x,t)$ and $a(x,t)$ for displacement, velocity and acceleration. In this setup, the nonlinear Green tensor can be replaced by

$$\varepsilon(u)(x,t) := \frac{\partial E}{\partial U}(0)(X,t) = \frac{1}{2}\big(\nabla_x u(x,t) + \nabla_x u(x,t)^T\big).$$

We note, that the linearized strain vanishes for *infinitesimal rigid body motions*

$$u(x,t) = \gamma(t)x + a(t),$$

where $a : [0,T] \to \mathbb{R}^d$ and $\gamma : [0,T] \to \mathbb{R}_{SKW}$ are smooth.

2.2.2 Balance equations

In the previous section, we introduced the basic quantities considered in continuum mechanics. In the following, we present the fundamental equations related to the motion of a body. We consider the balance of mass and momentum, and state the equations implied by these conservation laws. We do not give an insight into the derivation of these equations; we point the interested reader to monographs such as [MH94, FdV79, Cia88].

Let now $\Omega \subset \mathbb{R}^d$, $d = 2, 3$ be a smooth, connected domain. Let $\Phi : \Omega \times [0, T]$ be a motion of the body. We assume that all deformations are small, such that a linearization of the strain tensor is rectified. For $x \in \Omega, t \in [0, T]$, we introduce the quantities of

$$\text{mass density} \qquad \rho(x, t),$$
$$\text{specific force density} \qquad \bar{f}(x, t).$$

Let now $\omega \subset \Omega$ be some connected, smooth sub-domain of Ω. Let moreover $\omega_t := \Phi(\omega, t)$ be the deformed configuration of ω at time t. In any surface point $x \in \partial \omega$, we denote by n the outer unit normal vector. Then, we define the

$$\text{surface force density} \qquad \vec{t}(x, t, n),$$

which describes the forces acting in point x at time t onto a surface with normal n. If the balance of momentum below is satisfied, it can be shown that this vector-valued quantity depends linearly on the normal vector n. This statement is referred to as Cauchy's theorem, and proven e.g. in [MH94]. Then there exists a tensor field $\sigma(x, t)$ on Ω such that $\vec{t}(x, t, n) = \sigma(x, t)n$ on the boundary. This tensor field is called

$$\text{stress field} \qquad \sigma(x, t).$$

Conservation of mass Let ω be some sub-body of Ω as described above, and ω_t its deformed configuration at time t. Let ρ_R be the (prescribed) mass density of the reference configuration. Then, the law of conservation of mass states that

$$\int_\omega \rho_R(x)\, dx = \int_{\omega_t} \rho(x, t)\, dx \qquad \text{for } t \in (0, T]. \tag{2.1}$$

Using Reynold's Transport Theorem (Theorem 1.1 in Chapter 2 of [MH94]), one can show that having (2.1) for any $\omega \subset \Omega$ is equivalent to a partial differential equation, provided the mass density ρ and velocity v are smooth enough. Indeed, if both functions are differentiable, it is equivalent to impose

$$\frac{\partial \rho}{\partial t} = -\operatorname{div}(\rho v) \qquad \text{in } \Omega \times (0, T], \qquad (2.2)$$

$$\rho(0) = \rho_R \qquad \text{in } \Omega. \qquad (2.3)$$

Balance of momentum Under the same assumptions as above, Newton's second law states that for any smooth sub-domain ω the *balance of linear momentum* holds true,

$$\frac{\partial}{\partial t} \int_{\omega_t} \rho v \, dx = \int_{\partial \omega_t} \vec{t}(x, t, n) \, ds + \int_{\omega_t} \rho \bar{f} \, dx \qquad \text{for } t \in (0, T].$$

Similarly, the *balance of angular momentum* asserts that

$$\frac{\partial}{\partial t} \int_{\omega_t} x \times \rho v \, dx = \int_{\partial \omega_t} x \times \vec{t}(x, t, n) \, ds + \int_{\omega_t} x \times \rho \bar{f} \, dx \qquad \text{for } t \in (0, T].$$

Similar to the conservation of mass, one can deduce an equivalent formulation as a partial differential equation, given all involved fields are smooth enough. Then the two equations above imply *Cauchy's equation of motion* and the *symmetry of the stress tensor*

$$\rho \frac{\partial v}{\partial t} = \operatorname{div} \sigma + f \qquad \text{in } \Omega \times (0, T], \qquad (2.4)$$

$$\sigma = \sigma^T \qquad \text{in } \Omega \times (0, T]. \qquad (2.5)$$

Here, we use the volume force density $f := \rho \bar{f}$, and the characterization of $\vec{t}(x, t, n)$ by the stress tensor $\sigma(x, t)$.

Steady state In this thesis, we do not treat time-dependent problems. We rather consider steady state solutions, which do not change in time. We assume that all given quantities are independent of time t, and then compute the steady state displacement $u : \Omega \to \mathbb{R}^d$. Thus, both velocity and acceleration vanish. The equations of balance of momentum (2.4), (2.5) simplify to

$$0 = \operatorname{div} \sigma + f \qquad \text{in } \Omega, \qquad (2.6)$$

$$\sigma = \sigma^T \qquad \text{in } \Omega. \qquad (2.7)$$

If we also fix the mass density ρ to its initial value ρ_R, the material is called *incompressible*. Then, the equation of conservation of mass (2.2) reduces to

$$0 = \operatorname{div} u \qquad \text{in } \Omega. \qquad (2.8)$$

2.3 Material laws

So far, the equations of balance of momentum (2.6), (2.7) provided a relation between the stress tensor σ and the given volume forces f. In this section, we prescribe a material law, which states a linear relation between the strain and the stress tensor (Hooke's law).

Linear elastic materials A material is called *linear elastic*, if the stress tensor depends linearly on the strain. Then, there exists a fourth order tensor field $\bar{C}(x)$ such that

$$\sigma = \bar{C}\varepsilon(u) \quad \text{in } \Omega. \tag{2.9}$$

This relation is referred to as "Hooke's law". If $\bar{C}$ does not depend on x, the material is called *homogenous*. A material is *isotropic*, if it is independent of the direction of the main strains, i.e. of the eigenvectors of the strain tensor. One can show that a linear elastic, homogenous, isotropic material is determined by two constant values. One often uses the Young modulus $\bar{E}$ and the Poisson ratio $\bar{\nu}$ to characterize $\bar{C}$. Then, the relation between stress and strain reads

$$\sigma = \frac{\bar{E}}{1+\bar{\nu}}\left(\frac{\bar{\nu}}{1-2\bar{\nu}}\operatorname{tr}(\varepsilon(u))I + \varepsilon(u)\right). \tag{2.10}$$

Physically sensible values of $\bar{E}$ and $\bar{\nu}$ range between

$$0 < \bar{\nu} < 1/2 \quad \text{and} \quad \bar{E} > 0.$$

Another possibility is to use the Lamé constants $\bar{\lambda}$ and $\bar{\mu}$, which are related to $\bar{E}, \bar{\nu}$ via

$$\bar{\lambda} = \frac{\bar{E}\bar{\nu}}{(1+\bar{\nu})(1-2\bar{\nu})}, \quad \bar{\mu} = \frac{\bar{E}}{2(1+\bar{\nu})}.$$

Then, the stress-strain relation is defined by

$$\sigma = \bar{\lambda}\operatorname{tr}(\varepsilon(u))I + 2\bar{\mu}\varepsilon(u) = \bar{\lambda}\operatorname{div} u I + 2\bar{\mu}\varepsilon(u). \tag{2.11}$$

The fourth order tensor $\bar{C}$ exists and is invertible as long as $\bar{\lambda} < \infty$, or, equivalently, $\bar{\nu} < 1/2$. In this case, we may set

$$\bar{A} := \bar{C}^{-1}.$$

We call $\bar{A}$ the *compliance tensor*, its application to some stress function σ is given by

$$\bar{A}\sigma = \frac{1}{2\bar{\mu}}\operatorname{dev}\sigma + \frac{1}{d(d\bar{\lambda}+2\bar{\mu})}\operatorname{tr}(\sigma)I. \tag{2.12}$$

One sees from (2.11), that for $\bar{\lambda} \to \infty$, the divergence of the displacement field is penalized. In the limiting case, the divergence of u has to vanish. Having in mind equation (2.8), we see that this means that the material is incompressible. When using $\bar{E}$ and $\bar{\nu}$ as material parameters, incompressibility is reached for $\bar{\nu} = 1/2$. In this case, the fourth order tensor $\bar{C}$ does not exist. The compliance tensor $\bar{A}$ can still be defined in the incompressible limit, it is then singular, and its application to a stress function σ is given by

$$\bar{A}\sigma = \frac{1}{2\bar{\mu}} \operatorname{dev} \sigma.$$

Thus, the strain depends only on the deviator of the stress. We talk of *nearly incompressible materials*, if we mean that $\bar{\lambda} \to \infty$. In this case, straightforward analytical and numerical methods fail, one runs into stability troubles. Thus, special methods, which are stable with respect to $\bar{\lambda} \to \infty$ are needed. In the scope of this thesis, we will present a method which is feasible in this setup.

2.4 Boundary value problems

In this section, we add suitable conditions on the stress and displacement on the boundary of the body Ω. Later, we will see, that correctly chosen conditions ensure that the equations of elasticity presented above have a unique solution. From intuition it seems clear, that the deformation of and stresses occurring in the body are uniquely determined, if it is fixed along at least a part of its boundary, such that no rigid body motions can occur. If this is not the case, the solution can only be unique up to such rigid body motions, which do not affect the strains and stresses in the body.

Let now $\Omega \subset \mathbb{R}^d, d = 2, 3$ be an open, connected domain with suitably smooth boundary. Which smoothness assumptions on the boundary really are required will be discussed later in this thesis. We divide the boundary $\Gamma = \partial\Omega$ into two parts, $\Gamma = \Gamma_D \cup \Gamma_N$. These parts shall not intersect, $\Gamma_D \cap \Gamma_N = \emptyset$. There, Γ_D is regarded as closed set, while Γ_N is regarded as open. We further assume that Γ_D is non-trivial, i.e. of positive $d - 1$ dimensional measure $|\Gamma_D| > 0$. We include the case of $\Gamma_N = \emptyset$.

On the boundary part Γ_D, we impose *Dirichlet boundary conditions*, i.e. we fix the displacement u to some given surface displacement u_D,

$$u = u_D \quad \text{on } \Gamma_D.$$

The remaining part Γ_N is governed by *traction* or *Neumann boundary conditions*. There, the surface tractions $\vec{t}(x, n) = \sigma(x)n$ are prescribed,

$$\sigma n = \vec{t}_N \quad \text{on } \Gamma_N.$$

There are more physically sensible types of boundary conditions, such as the *pressure boundary condition*, where the pressure is prescribed. Then, the surface traction points in normal direction, and the tangential tractions vanish. This is obtained setting

$$\sigma n = pn.$$

However, we do not treat this kind of condition separately, it can be viewed within the more general framework of traction boundary conditions.

Chapter 3

Variational Framework

In this chapter, we present several variational formulations for the elasticity problem on a suitable, bounded and connected domain $\Omega \subset \mathbb{R}^d$, $d = 2, 3$.

Problem 3.1 (Elasticity problem). *Find $u : \Omega \to \mathbb{R}^d$, and $\sigma : \Omega \to \mathbb{R}^{d \times d}$ symmetric such that*

$$A\sigma - \varepsilon(u) = 0 \quad \text{in } \Omega, \tag{3.1}$$
$$-\operatorname{div}(\sigma) = f \quad \text{in } \Omega, \tag{3.2}$$

with boundary conditions

$$u = u_D \quad \text{on } \Gamma_D, \tag{3.3}$$
$$\sigma_n = \vec{t}_N \quad \text{on } \Gamma_N. \tag{3.4}$$

We introduce the scalar-valued function space $H^1(\Omega)$, as well as the vector-valued spaces $H(\operatorname{curl}; \Omega)$ and $H(\operatorname{div}; \Omega)$. As the stress is a tensor-valued symmetric quantity, we discuss feasible spaces and some of their properties for such fields. Then, we present three well-known variational formulations for the elasticity problem. The first is the primal formulation, where all derivatives are applied to the displacement. The second one is the mixed formulation by Hellinger and Reissner. Here, the displacement can be totally discontinuous, but the divergence of the symmetric stress tensor must exist in a weak sense. The third is a mixed formulation with weak symmetry, where, as the name suggests, the symmetry of the stress tensor is imposed via Lagrangian multipliers.

In the second part of this chapter, we introduce a new mixed variational formulation, which lies in between the first two concepts. We search for the displacement in $H(\operatorname{curl}; \Omega)$, which lies in between the primal and the Hellinger-Reissner choice. We provide a proper space $H(\operatorname{div}\operatorname{div}; \Omega)$ for the stress. We show existence and uniqueness for all the different formulations, using abstract theory for mixed problems.

This chapter is organized as follows: In Section 3.1, spaces allowing for weak gradient, curl, and divergence operators on domains and manifolds are introduced, and their basic properties are recalled. In Section 3.2, abstract theory for mixed variational problems is presented, and well-known mixed formulations of the elasticity problem are shortly described. Section 3.3 deals with the formulation and analysis of our new mixed approach.

3.1 Function spaces

In the following, we define differential operators and function spaces on an exemplary domain D. Throughout this work, we will assume that any domain satisfies the conditions below.

Assumption 3.2. *Let $D \subset \mathbb{R}^d, d = 2, 3$ be an open Lipschitz domain, i.e. its boundary can be represented by a finite number of Lipschitz continuous functions. We assume that the boundary ∂D is*

- *either smooth, $\partial D \in \mathcal{C}^\infty$,*
- *or polygonal/polyhedral, consisting of straight open facets $\Gamma_i, i = 1, \ldots, s$, such that*

$$\partial D = \bigcup_{i=1}^{s} \bar{\Gamma}_i, \qquad \Gamma_i \cap \Gamma_j = \emptyset, \qquad n_i := n|_{\Gamma_i} = const.$$

In the following, we will call the boundary polyhedral, no matter whether in two or three space dimensions. Note that, in the plane, the "facets" Γ_i are the edges of the polygon, whereas facets in three-dimensional space are faces of a polyhedron. We will meet the concept of facets again when finite element triangulations of the domain are concerned.

In this section, let D be of unit size, $\operatorname{diam} D \simeq 1$. Then, all constants depending on D only depend on the shape of D, and not on its size; proper scalings of terms can be introduced when needed.

The differential operators from Section 2.1 are defined for smooth functions. Let now

$$L^2(D) := \left\{ w : D \to \mathbb{R} \ \Big| \ \int_D |w|^2 \, dx < \infty \right\}$$

be the standard Lebesgue space of square integrable functions. It is equipped with the following inner product and induced norm

$$(u, v)_{L^2(D)} := (u, v)_D := \int_D uv \, dx, \qquad \|u\|_{L^2(D)} := \|u\|_D := \sqrt{(u, u)_D}.$$

The gradient, curl and divergence operators can also be defined in weak sense:

Definition 3.3. *For $w \in L^2(D)$, $g = \nabla w \in [L^2(D)]^d$ is called* weak gradient *of w iff*

$$\int_D g \cdot \phi \, dx = -\int_D w \operatorname{div} \phi \, dx \qquad \forall \phi \in [\mathcal{C}_0^\infty(D)]^d.$$

For $v \in [L^2(D)]^d$, $c = \operatorname{curl} v \in [L^2(D)]^{d(d-1)/2}$ is called weak curl of v iff

$$\int_D c \cdot \phi \, dx = \int_D v \cdot \operatorname{Curl} \phi \, dx \qquad \forall \phi \in [\mathcal{C}_0^\infty(D)]^{d(d-1)/2}.$$

For $q \in [L^2(D)]^d$, $d = \operatorname{div} q \in L^2(D)$ is called weak divergence of q iff

$$\int_D d \cdot \phi \, dx = -\int_D q \cdot \nabla \phi \, dx \qquad \forall \phi \in \mathcal{C}_0^\infty(D).$$

3.1.1 Sobolev spaces on a domain

The weak differential operators define Hilbert spaces, which are appropriate for the analysis of PDEs in variational form.

Definition 3.4. *Define the Sobolev space $H^1(D)$, and the spaces $H(\operatorname{curl}; D), H(\operatorname{div}; D)$ allowing for weak curl- and divergence operators,*

$$\begin{aligned} H^1(D) &:= \{w \in L^2(D) : \nabla w \in [L^2(D)]^d\}, \\ H(\operatorname{curl}; D) &:= \{v \in [L^2(D)]^d : \operatorname{curl} v \in [L^2(D)]^{d(d-1)/2}\}, \\ H(\operatorname{div}; D) &:= \{q \in [L^2(D)]^d : \operatorname{div} q \in L^2(D)\}. \end{aligned}$$

When equipped with inner products

$$\begin{aligned} (u, v)_{H^1(D)} &:= (u, v)_D + (\nabla u, \nabla v)_D, \\ (u, v)_{H(\operatorname{curl}; D)} &:= (u, v)_D + (\operatorname{curl} u, \operatorname{curl} v)_D, \\ (q, p)_{H(\operatorname{div}; D)} &:= (q, p)_D + (\operatorname{div} q, \operatorname{div} p)_D, \end{aligned}$$

and the induced norms $\|\cdot\|_{H^1(D)}$, $\|\cdot\|_{H(\operatorname{curl};D)}$ and $\|\cdot\|_{H(\operatorname{div};D)}$, they are Hilbert spaces.

Sobolev spaces of higher order can be defined similarly, for $k \in \mathbb{N}$ we set

$$\begin{aligned} H^k(D) &:= \{w \in L^2(D) : \nabla^k w \in [L^2(D)]^{d^k}\}, \\ H^k(\operatorname{curl}; D) &:= \{v \in [H^k(D)]^d : \operatorname{curl} v \in [H^k(D)]^{d(d-1)/2}\}, \\ H^k(\operatorname{div}; D) &:= \{q \in [H^k(D)]^d : \operatorname{div} q \in H^k(D)\}. \end{aligned}$$

When $k = 0$, we have $H^0(D) = L^2(D)$. For the space $H^k(D)$, we use the following semi-norm and norm

$$|w|_{H^k(D)} := \|\nabla^k w\|_{L^2(D)}, \qquad \|w\|_{H^k(D)}^2 := \|w\|_{L^2(D)}^2 + |w|_{H^k(D)}^2.$$

The Sobolev spaces $H^s(D)$ with real index $s > 0$ can be defined using the Fourier transform; see e.g. [GR86]. The spaces $H^s(\operatorname{curl}; D)$ and $H^s(\operatorname{div}; D)$ can be defined accordingly. It can be shown, that the spaces $H^1(D)$, $H(\operatorname{curl};D)$, and $H(\operatorname{div};D)$ are the closure of $\mathcal{C}^\infty(\bar{D})$ in the respective norms.

Let now $A \subset \partial D$ be a part of the boundary of the domain D. We define $\mathcal{C}^k_{0,A}(D)$ to be the subset of $\mathcal{C}^k(D)$ consisting of functions whose trace and first $k-1$ derivatives vanish on the boundary part A. By $\mathcal{C}^\infty_{0,A}(D)$, we denote the set of infinitely often continuously differentiable functions with compact support on $\bar{D}\backslash A$. Then, we define

$$\begin{aligned}
H^k_{0,A}(D) &:= \overline{\mathcal{C}^\infty_{0,A}(D)}^{\|\cdot\|_{H^k(D)}}, & k \in \mathbb{N}, \\
H_{0,A}(\operatorname{curl};D) &:= \overline{[\mathcal{C}^\infty_{0,A}(D)]^d}^{\|\cdot\|_{H(\operatorname{curl};D)}}, \\
H_{0,A}(\operatorname{div};D) &:= \overline{[\mathcal{C}^\infty_{0,A}(D)]^d}^{\|\cdot\|_{H(\operatorname{div};D)}}.
\end{aligned}$$

We write $H^k_0(D)$, $H_0(\operatorname{curl};D)$ and $H_0(\operatorname{div};D)$ in case of $A = \partial D$. These spaces satisfy certain kinds of homogenous boundary conditions on A, which we will see in more detail in Section 3.1.3. We can now define distributional Sobolev spaces of negative index

$$H^{-k}(D) := H^k_0(D)^*, \qquad H^{-k}_A(D) := H^k_{0,A}(D)^*, \qquad k \in \mathbb{N}.$$

These duality relations imply to use the norm

$$\|q\|_{H^{-k}_A(D)} := \sup_{v \in H^k_{0,A}(D)} \frac{\langle q, v \rangle_{H^k(D)}}{\|v\|_{H^k(D)}}.$$

We denote the corresponding inner product by $(\cdot,\cdot)_{H^{-k}_A(D)}$.

We introduce the Lebesgue spaces of symmetric and skew-symmetric, tensor-valued functions

$$\begin{aligned}
L^2_{SYM}(D) &:= \{\tau \in [L^2(D)]^{d\times d} : \tau = \tau^T\}, \\
L^2_{SKW}(D) &:= \{\tau \in [L^2(D)]^{d\times d} : \tau = -\tau^T\}.
\end{aligned}$$

Analogously, the space of symmetric tensor fields with weak divergence is

$$H_{SYM}(\operatorname{div};D) := \{\tau \in L^2_{SYM}(D) : \operatorname{div} \tau \in [L^2(\operatorname{div};D)]^d\}.$$

As their scalar- and vector-valued equivalents, these spaces are Hilbert spaces.

3.1.2 Sobolev spaces on a manifold

We again consider a domain D satisfying Assumption 3.2. Let $A \subseteq \partial D$ denote a bounded, Lipschitzian manifold of dimension $d-1$. Let moreover $B := \partial D \backslash A$ be its complement on the boundary. Note that also the case $A = \partial D$ is included.

We can define the Lebesgue space $L^2(A)$ as the space of square integrable functions on this manifold. We will heavily use the space

$$H^{1/2}(A) := \overline{C^\infty(\bar{A})}^{\|\cdot\|_{H^{1/2}(A)}}.$$

Here, $\|\cdot\|_{H^{1/2}(A)}$ is the norm induced by the inner product

$$(u,v)_{H^{1/2}(A)} := \int_A \int_A \frac{(u(x)-u(y))(v(x)-v(y))}{|x-y|^d} ds_x ds_y + \int_A uv\, ds_x.$$

This definition is equivalent to one using the Fourier transform, as e.g. given in [GR86]. If $A \neq \partial D$, the space

$$H^{1/2}_{00}(A) := \overline{C_0^\infty(A)}^{\|\cdot\|_{H^{1/2}(\partial D)}}$$

can be extended by zero to the whole boundary ∂D, and the extension lies in $H^{1/2}(\partial D)$. We see that $H^{1/2}_{00}(\partial D) := H^{1/2}(\partial D)$. Following [BC01a], we introduce the dual spaces

$$H^{-1/2}(A) := [H^{1/2}(A)]^*, \qquad H^{-1/2}_{00}(A) := [H^{1/2}_{00}(A)]^*.$$

The space $H^{1/2}(\partial D)$ is scalar-valued. Next, we define vector-valued spaces on the surface. Therefore, on a polyhedral boundary consisting of facets $\{\Gamma_i\}_{i=1,\dots,s}$ as described in Assumption 3.2, we need to introduce the space of functions piecewise in $H^{1/2}(\partial D)$,

$$H^{1/2}_{pw}(\partial D) := \left\{ w \in L^2(\partial D) : w|_{\Gamma_i} \in H^{1/2}(\Gamma_i) \right\},$$

with norm

$$\|w\|_{H^{1/2}_{pw}(\partial D)} := \left(\sum_{i=1}^s \|w\|^2_{H^{1/2}(\Gamma_i)} \right)^{1/2}.$$

In case of a smooth boundary, we simply set $H^{1/2}_{pw}(\partial D) := H^{1/2}(\partial D)$. Of course, this space can be restricted to $H^{1/2}_{pw}(A)$ directly. Let us now define

$$\begin{aligned} L^2_\tau(\partial D) &:= \{ q \in [L^2(\partial D)]^d : q \cdot n = 0 \text{ a.e. on } \partial D \}, \\ H^{1/2}_\tau(\partial D) &:= \{ q \in L^2_\tau(\partial D) : q \in [H^{1/2}_{pw}]^d \}. \end{aligned}$$

The first space is the vector-valued Lebesgue space of square integrable functions, which lie in the tangential plane of ∂D almost everywhere. The second space, $H^{1/2}_\tau$, consists of functions in the tangential plane whose components are piecewise in $H^{1/2}$. Now, we provide a space, which, as we will see later, is a proper trace space for the tangential component of $[H^1(D)]^d$. For a smooth boundary, we already find this space in $H^{1/2}_\tau$. Now, we introduce such a space for a polyhedral boundary consisting of facets $\{\Gamma_i\}_{i=1,\dots,s}$. We first do so for the case of three space dimensions. All following definitions are taken from [BC01a, BC01b]. Let $E_{ij} := \bar{\Gamma}_i \cap \bar{\Gamma}_j$ be the edge between facets Γ_i, Γ_j, and let τ_{ij} be the tangential vector along this edge. On this edge, let $\tau_i := \tau_{ij} \times n_i$ be the vector lying in Γ_i, and orthogonal to the edge. We define

$$H_\|^{1/2}(\partial D) := \left\{q \in H_\tau^{1/2}(\partial D) : q|_{\Gamma_i} \cdot \tau_{ij} \stackrel{1/2}{=} q|_{\Gamma_j} \cdot \tau_{ij} \text{ on all } E_{ij}\right\},$$
$$H_\perp^{1/2}(\partial D) := \left\{q \in H_\tau^{1/2}(\partial D) : q|_{\Gamma_i} \cdot \tau_i \stackrel{1/2}{=} q|_{\Gamma_j} \cdot \tau_j \text{ on all } E_{ij}\right\}.$$

There, $\stackrel{1/2}{=}$ on an edge E_{ij} means

$$q_i \stackrel{1/2}{=} q_j \text{ on } E_{ij} \quad \text{iff} \quad \int_{\Gamma_i}\int_{\Gamma_j} \frac{|q_i(x) - q_j(y)|^2}{|x-y|^3} ds_x ds_y < \infty.$$

Intuitively, this means, that functions in $H_\|^{1/2}(\partial D)$ have continuous tangential components in the surface, as well as functions in $H_\perp^{1/2}(\partial D)$ have continuous normal components.

For $D \subset \mathbb{R}^2$ with polyhedral boundary, let V_{ij} be the vertex between facets Γ_i and Γ_j. Similarly to the 3d case, let n_i be the normal to facet Γ_i, and let $\tau_i = n_i^\perp$ be the tangent vector. There exists no equivalent to the edge tangent vector τ_{ij}. The spaces $H_\perp^{1/2}(\partial D)$ and $H_\|^{1/2}(\partial D)$ are then defined via

$$H_\|^{1/2}(\partial D) := H_\tau^{1/2}(\partial D),$$
$$H_\perp^{1/2}(\partial D) := \left\{q \in H_\tau^{1/2}(\partial D) : q|_{\Gamma_i} \cdot \tau_i \stackrel{1/2}{=} q|_{\Gamma_j} \cdot \tau_j \text{ on } V_{ij}\right\} \quad \text{for } d = 2,$$

where $\stackrel{1/2}{=}$ on the vertex V_{ij} is defined in a similar way as for an edge in three dimensions,

$$q_i \stackrel{1/2}{=} q_j \text{ on } V_{ij} \quad \text{iff} \quad \int_{\Gamma_i}\int_{\Gamma_j} \frac{|q_i(x) - q_j(y)|^2}{|x-y|^2} ds_x ds_y < \infty.$$

If the boundary ∂D is smooth, we set $H_\|^{1/2}(\partial D) := H_\perp^{1/2}(\partial D) := H_\tau^{1/2}(\partial \Omega)$ in both two and three dimensions. We set $H_\|^{-1/2}(\partial D) := [H_\|^{1/2}(\partial D)]^*$, and $H_\perp^{-1/2}(\partial D) := [H_\perp^{1/2}(\partial D)]^*$.

Now, we introduce even less regular vector-valued spaces on the surface. We do so for the case of three space dimensions first. All following definitions are taken from [BC01a, BC01b]. We need differential operators living on ∂D: We use $\nabla_{\partial D}$, $\text{curl}_{\partial D}$ and $\text{div}_{\partial D}$, which are defined for smooth domains e.g. in [Mon03, Section 3.4], for polyhedral domains in [BC01a]. We define

$$H_\perp^{-1/2}(\text{curl}_{\partial D}; \partial D) := \{q \in H_\perp^{-1/2}(\partial D) : \text{curl}_{\partial D}\, q \in H^{-1/2}(\partial D)\},$$
$$H_\|^{-1/2}(\text{div}_{\partial D}; \partial D) := \{q \in H_\|^{-1/2}(\partial D) : \text{div}_{\partial D}\, q \in H^{-1/2}(\partial D)\}.$$

In two space dimensions, the boundary is of local dimension one, therefore the notion of gradient and divergence is replaced by the tangential derivative. The curl operator does not exist.

Nevertheless, to unify notation in two and three space dimensions, we define for $D \subset \mathbb{R}^2$

$$H_\perp^{-1/2}(\operatorname{curl}_{\partial D}; \partial D) := H_\perp^{-1/2}(\partial D),$$
$$H_\|^{-1/2}(\operatorname{div}_{\partial D}; \partial D) := H_\perp^{1/2}(\partial D).$$

Later on, we will obtain the same trace theorems in both cases $d = 2, 3$. Note that, also in both two and three dimensions, the surface spaces are dual to each other, we have [BC01b]

$$H_\perp^{-1/2}(\operatorname{curl}_{\partial D}; \partial D) = \left[H_\|^{-1/2}(\operatorname{div}_{\partial D}; \partial D)\right]^*.$$

All these spaces can be restricted to a part A of the boundary, we write

$$H_\perp^{-1/2}(\operatorname{curl}_A; A) \quad \text{and} \quad H_\|^{-1/2}(\operatorname{div}_A; A).$$

If A has a boundary ∂A, it is possible to define subspaces

$$H_\perp^{-1/2}(\operatorname{curl}_A^0; A) \quad \text{and} \quad H_\|^{-1/2}(\operatorname{div}_A^0; A)$$

such that functions can be extended conformingly by zero to the whole boundary ∂D.

Well aware that spaces with negative index contain distributions, we call them function spaces. We may even refer to their elements as functions, when it seems more intuitive to do so.

Last, we need to introduce one further space on a manifold, namely the Neumann trace space of $H^2(D) \cap H_0^1(D)$. We set

$$H_{n,0}^{1/2}(\partial D) := \left\{ \frac{\partial \varphi}{\partial n} : \varphi \in H^2(D) \cap H_0^1(D) \right\}, \tag{3.5}$$

with its norm defined by

$$\|q\|_{H_{n,0}^{1/2}(\partial D)} := \inf_{\substack{\varphi \in H^2 \cap H_0^1 \\ \partial \varphi / \partial n = q}} \|w\|_{H^2(D)}. \tag{3.6}$$

Note that, up to now, we did not introduce the notion of trace operators formally. In Section 3.1.3, we will rectify this choice. For a smooth boundary ∂D, this space is equal to $H^{1/2}(\partial D)$. In case of a polyhedral domain with boundary facets $\{\Gamma_i\}_{i=1...s}$, this space consists of piecewise $H_{00}^{1/2}$ contributions,

$$H_{n,0}^{1/2}(\partial D) = \{ w \in L^2(\partial D) : w|_{\Gamma_i} \in H_{00}^{1/2}(\Gamma_i) \}.$$

This characterization can be derived from the more involved setup in [Gri85, Theorem 1.5.8.2]. For the case of mixed boundary conditions, we also introduce

$$H_{n,0,A}^{1/2}(\partial D) := \left\{ \frac{\partial \varphi}{\partial n} : \phi \in H^2(D) \cap H_{0,A}^1(D) \right\}. \tag{3.7}$$

We denote the dual spaces

$$H_n^{-1/2}(\partial D) := H_{n,0}^{1/2}(\partial D)^*, \tag{3.8}$$

$$H_{n,A}^{-1/2}(\partial D) := H_{n,0,A}^{1/2}(\partial D)^*. \tag{3.9}$$

3.1.3 Trace theorems

The spaces $H^1(D), H(\text{curl}; D)$, and $H(\text{div}; D)$ allow different types of boundary conditions. We provide trace and inverse trace theorems for all spaces. From these theorems, we deduce interface conditions for piecewise defined vector fields.

3.1.3.1 Traces in the classical sense

For a continuous function, the trace operator

$$\text{tr}_{\partial D} : \mathcal{C}(\bar{D}) \to \mathcal{C}(\partial D), \qquad (\text{tr}_{\partial D} w)(x) := w(x) \text{ for } x \in \partial D$$

is well defined on a Lipschitz domain. Note that, on the boundary ∂D of a Lipschitz domain, the outward unit normal vector n can be defined almost everywhere. For a vector $v \in \mathbb{R}^d$, its normal and tangential components with respect to n are given by

$$v_n = v \cdot n, \qquad v_\tau = v - v_n n.$$

In three space dimensions, there holds $v_\tau = n \times (v \times n)$. Thus, for a vector valued, continuous function v, we may also define the normal and tangential traces $\text{tr}_{\partial D,n} v, \text{tr}_{\partial D,\tau} v$ almost everywhere by

$$\text{tr}_{\partial D,n} v = v_n, \qquad \text{tr}_{\partial D,\tau} v = v_\tau.$$

Note that $\text{tr}_{\partial D,n} v$ is a scalar quantity, whereas $\text{tr}_{\partial D,\tau} v$ is vector-valued and lies in the tangential space of ∂D, which is locally of dimension $d-1$. For $A \subset \partial D$, trace operators $\text{tr}_A, \text{tr}_{A,\tau}$, and $\text{tr}_{A,n}$ can be defined as the restriction of the respective traces to A.

We will see that the notion of a trace operator can be extended to H^1 functions, whereas functions in $H(\text{curl})$ allow for a tangential trace, as well as functions in $H(\text{div})$ have a normal trace.

3.1.3.2 Traces of the space H^1

The classical trace operator can be extended to the Sobolev space $H^1(D)$; its image is then $H^{1/2}(\partial D)$. Moreover, any function w in the trace space $H^{1/2}(\partial D)$ can be extended to some $\tilde{w} \in H^1(D)$, and both the trace and extension operator are continuous. For the proof, we refer to [Gri85].

Theorem 3.5 (Trace theorem for H^1). *Let $D \subset \mathbb{R}^d, d = 2,3$ be a bounded Lipschitz domain satisfying Assumption 3.2, $\operatorname{diam} D \simeq 1$,*

1. *The trace operator $\operatorname{tr}_{\partial D}$ is well defined on $H^1(D)$ as an extension from $C^\infty(\bar{D})$. It is continuous from $H^1(D)$ to $H^{1/2}(\partial D)$; there exists a constant $c_{tr} > 0$ such that*

$$\|\operatorname{tr}_{\partial D} u\|_{H^{1/2}(\partial D)} \leq c_{tr} \|u\|_{H^1(D)} \qquad \forall u \in H^1(D).$$

2. *For $g \in H^{1/2}(\partial D)$, there exists some $u \in H^1(D)$ such that*

$$\operatorname{tr}_{\partial D} u = g \text{ on } \partial D \quad \text{and} \quad \|u\|_{H^1(D)} \leq c_{ext} \|g\|_{H^{1/2}(\partial D)}.$$

The constants c_{tr}, c_{ext} are positive and only depend on the shape of the domain D.

The second statement is often referred to as *inverse trace theorem* or *extension theorem*. Both directions are needed for the incorporation of boundary conditions in the primal variational setting.

There exists also a trace theorem for the vector-valued space $[H^1(D)]^d$. For smooth domains, it follows directly from Theorem 3.5, for polyhedral boundaries we refer to [BC01a].

Theorem 3.6 (Tangential trace of $(H^1)^d$). *Let $D \subset \mathbb{R}^d, d = 2,3$ be a bounded Lipschitz domain satisfying Assumption 3.2, $\operatorname{diam} D \simeq 1$,*

1. *The tangential trace operator $\operatorname{tr}_{\tau,\partial D}$ is continuous from $[H^1(D)]^d$ to $H_\|^{1/2}(\partial D)$, there exists a constant $c_{tr,\tau} > 0$ such that*

$$\|\operatorname{tr}_{\tau,\partial D} u\|_{H_\|^{1/2}(\partial D)} \leq c_{tr,\tau} \|u\|_{H^1(D)} \qquad \forall u \in [H^1(D)]^d.$$

2. *For $g \in H_\|^{1/2}(\partial D)$, there exists some $u \in [H^1(D)]^d$ such that*

$$\operatorname{tr}_{\tau,\partial D} u = g \text{ on } \partial D \quad \text{and} \quad \|u\|_{H^1(D)} \leq c_{ext,\tau} \|g\|_{H_\|^{1/2}(\partial D)}.$$

The constants $c_{\tau,tr}, c_{\tau,ext}$ are positive and only depend on the shape of the domain D.

Remark 3.7. Note that, in three dimensions, this statement is equivalent to posing continuity and surjectivity of the rotated tangential trace $u \mapsto u \times n$ from $[H^1(D)]$ to $H_\perp^{1/2}(\partial D)$.

3.1.3.3 Traces of the space $H(\operatorname{curl})$

The space $H(\operatorname{curl}; D)$ allows the definition of a tangential trace operator, which coincides with the tangential trace in the classical sense for smooth functions. The following theorem holds by [BC01b].

Theorem 3.8 (Trace theorem for $H(\text{curl})$). *Let $D \subset \mathbb{R}^d, d = 2, 3$ be a bounded Lipschitz domain as in Assumption 3.2, $\operatorname{diam} D \simeq 1$.*

1. *The classical tangential trace operator $\operatorname{tr}_{\partial D, \tau}$ can be extended to $H(\text{curl}; D)$. It is continuous from $H(\text{curl}; D)$ to $H_\perp^{-1/2}(\text{curl}_{\partial D}; \partial D)$, there exists a constant $c_{\text{tr}_\tau} > 0$ such that*

$$\| \operatorname{tr}_{\partial D, \tau} v \|_{H_\perp^{-1/2}(\text{curl}_{\partial D}; \partial D)} \le c_{\text{tr}_\tau} \| v \|_{H(\text{curl}; D)} \qquad \forall v \in H(\text{curl}; D).$$

2. *Any $g \in H_\perp^{-1/2}(\text{curl}_{\partial D}; \partial D)$ can be extended to $v \in H(\text{curl}; D)$ such that*

$$\operatorname{tr}_{\partial D, \tau} v = g \text{ on } \partial D \quad \text{and} \quad \| v \|_{H(\text{curl}; D)} \le c_{\text{ext}_\tau} \| g \|_{H_\perp^{-1/2}(\text{curl}_{\partial D}; \partial D)}.$$

The constants $c_{\text{tr}_\tau}, c_{\text{ext}_\tau}$ are positive and depend only on the shape of the domain D.

Remark 3.9. *In three dimensions, we observe similarly to Remark 3.7, that we equivalently have the rotated tangential trace operator $u \mapsto u \times n$ continuous and surjective from $H(\text{curl}; D)$ to $H_\parallel^{-1/2}(\text{div}_{\partial D}; \partial D)$.*

Theorem 3.8 is necessary for the analysis of boundary conditions for the tangential trace of a function in $H(\text{curl})$. We will see later, that these conditions are essential in our mixed variational formulation, while conditions on the normal trace can be imposed only weakly, and are therefore natural.

3.1.3.4 Traces of the space $H(\text{div})$

The normal trace operator $\operatorname{tr}_{\partial D, n}$ can be extended from $[\mathcal{C}^\infty(\bar{D})]^d$ to $H(\text{div}; D)$. The mapping is continuous and surjective onto the space $H^{-1/2}(\partial D)$, as the next theorem states. A proof for the statement is given in [GR86, Corollary 2.8].

Theorem 3.10 (Trace theorem for $H(\text{div})$). *Let $D \subset \mathbb{R}^d, d = 2, 3$ be a bounded Lipschitz domain satisfying Assumption 3.2, $\operatorname{diam} D \simeq 1$.*

1. *The trace operator $\operatorname{tr}_{\partial D, n}$ is well defined on $H(\text{div}; D)$ as an extension from $[\mathcal{C}^\infty(\bar{D})]^d$. It is continuous onto $H^{-1/2}(\partial D)$, there exists a constant $c_{\text{tr}_n} > 0$ such that*

$$\| \operatorname{tr}_{\partial D, n} q \|_{H^{-1/2}(\partial D)} \le c_{\text{tr}_n} \| q \|_{H(\text{div}; D)} \qquad \forall q \in H(\text{div}; D).$$

2. *For $g \in H^{-1/2}(\partial D)$, there exists some $q \in H(\text{div}; D)$ such that*

$$\operatorname{tr}_{\partial D, n} q = g \text{ on } \partial D \quad \text{and} \quad \| q \|_{H(\text{div}; D)} \le c_{\text{ext}_n} \| g \|_{H^{-1/2}(\partial D)}.$$

The constants $c_{\text{tr}_n}, c_{\text{ext}_n}$ are positive and depend only on the shape of the domain D.

We will see that boundary conditions on the normal component of a function in $H(\text{div})$ are essential, and have to be included into the solution space of a variational formulation.

Boundary conditions Let again $A \subset \partial D$ be a part of the boundary. It can be shown that the spaces $H^1_{0,A}(D), H_{0,A}(\text{curl}; D)$ and $H_{0,A}(\text{div}; D)$ satisfy the respective essential homogenous boundary conditions on A,

$$\begin{aligned} H^1_{0,A}(D) &= \{w \in H^1(D) : \text{tr}_A\, w = 0\}, \\ H_{0,A}(\text{curl}; D) &= \{v \in H(\text{curl}; D) : \text{tr}_{A,\tau}\, v = 0\}, \\ H_{0,A}(\text{div}; D) &= \{q \in H(\text{div}; D) : \text{tr}_{A,n}\, q = 0\}. \end{aligned}$$

Functions in $H^1_0(D), H_0(\text{curl}; D)$, and $H_0(\text{div}; D)$ can be extended to $H^1(\mathbb{R}^d), H(\text{curl}; \mathbb{R}^d)$ and $H(\text{div}; \mathbb{R}^d)$ by zero.

In an abuse of notation, we will often write $u|_{\partial D}$ meaning $\text{tr}_{\partial D}\, u$ for $u \in H^1(D)$, or similarly $v_\tau|_{\partial D}$ and $q_n|_{\partial D}$ meaning the respective traces $\text{tr}_{\partial D, \tau}\, v, \text{tr}_{\partial D, n}\, q$ of $v \in H(\text{curl}; D)$ and $q \in H(\text{div}; D)$. When clear from the context that the evaluation is on the boundary ∂D, we may even use u, v_τ and q_n only.

3.1.4 Green's formulae

We recall some standard facts on integration by parts for the operators ∇, curl, and div. These identities are well-known in the classical sense, but also hold in the weak setting, as long as all arising integrals are well-defined at least in the sense of duality products. For boundary integrals, the respective trace operators have to be used. In order not to complicate notation, we write integrals in any case, well aware that duality products are necessary when considering traces in Sobolev spaces of negative indices.

Theorem 3.11 (Green's formulae). *The following integration by parts formulae hold on a Lipschitz domain $D \subset \mathbb{R}^d, d = 2, 3$:*

$$\int_D \nabla u \cdot q\, dx = -\int_D u \,\text{div}\, q\, dx + \int_{\partial D} u\, q_n\, ds \qquad \forall u \in H^1(D), q \in H(\text{div}; D), \quad (3.10)$$

$$\int_D \text{curl}\, u \cdot v\, dx = \int_D u \cdot \text{Curl}\, v\, dx - \int_{\partial D} (u \times n) \cdot v\, ds$$
$$\forall u \in H(\text{curl}; D), v \in [H^1(D)]^{d(d-1)/2}, \quad (3.11)$$

$$\int_D \varepsilon(u) : \tau\, dx = -\int_D u \cdot \text{div}\, \tau\, dx - \int_{\partial D} u \cdot \tau_n\, ds$$
$$\forall u \in [H^1(D)]^d, \tau \in H_{\text{SYM}}(\text{div}; D). \quad (3.12)$$

Here, we set $u \times n := u_2 n_1 - u_1 n_2$ in formula (3.11) in case of $d = 2$. Note that identity (3.12) can be seen directly from equation (3.10) and the symmetry of all involved tensor-valued quantities.

This theorem can be used to show interface conditions for piecewise defined functions.

Theorem 3.12 (Interface conditions). *Let $\mathcal{D} = \{D_1, \ldots, D_N\}$ be a non-overlapping domain decomposition for D, i.e. $D_i \cap D_j = \emptyset$ for $i \neq j \in \{1, \ldots, N\}$ and $\bar{D} = \bigcup \bar{D}_i$. Let $\Gamma_{ij} := \partial D_i \cap \partial D_j$ be the common interface of any two sub-domains $D_i, D_j \in \mathcal{D}$. Let $u, v,$ and q be defined piecewise such that $u_i := u|_{D_i} \in H^1(D_i)$, $v_i := v|_{D_i} \in H(\text{curl}; D_i)$ and $q_i := q|_{D_i} \in H(\text{div}; D_i)$ for $i = 1, \ldots N$. Then there holds*

- *if $\text{tr}_{\Gamma_{ij}} u_i = \text{tr}_{\Gamma_{ij}} u_j$ for $i, j = 1, \ldots, N$, u lies in the global space $H^1(D)$, and $(\nabla u)|_{D_i} = \nabla(u_i)$.*

- *if $\text{tr}_{\Gamma_{ij}, \tau} v_i = \text{tr}_{\Gamma_{ij}, \tau} v_j$ for $i, j = 1, \ldots, N$, v lies in the global space $H(\text{curl}; D)$, and $(\text{curl } v)|_{D_i} = \text{curl}(v_i)$.*

- *if $\text{tr}_{\Gamma_{ij}, n} q_i = \text{tr}_{\Gamma_{ij}, n} q_j$ for $i, j = 1, \ldots, N$, q lies in the global space $H(\text{div}; D)$, and $(\text{div } q)|_{D_i} = \text{div}(q_i)$.*

From this theorem one can see that, when constructing finite element functions using piecewise polynomials on some prescribed mesh, one has to keep continuity across interfaces to be conforming for H^1. Similarly, tangential continuity for $H(\text{curl})$ and normal continuity for $H(\text{div})$ have to be satisfied. Boundary conditions of the respective types have to be included into solution spaces of a variational formulation or finite element discretization, and are therefore essential.

3.1.5 Dual space of $H(\text{curl})$

Let again $A \subset \partial D$ be a part of the boundary. We investigate the dual space of $H_{0,A}(\text{curl}; D)$. This includes $[H(\text{curl}; D)]^*$, $[H_0(\text{curl}; D)]^*$ by setting either $A = \emptyset$ or $A = \partial D$. The following lemma states, that the dual space we are looking for is a distributional space allowing for a distributional divergence.

Lemma 3.13. *Let $D \subset \mathbb{R}^d, d = 2, 3$ be a bounded, connected Lipschitz domain as described in Assumption 3.2, and let $A \subset \partial D$ be a part of the boundary. Then*

$$[H_{0,A}(\text{curl}; D)]^* = H_A^{-1}(\text{div}; D). \tag{3.13}$$

where $H_A^{-1}(\text{div}; D) := \{q \in [H_A^{-1}(D)]^d : \text{div } q \in H_A^{-1}(D)\}$ is the space of all distributions from $[H_A^{-1}(D)]^d$, such that the distributional divergence, in this context defined by

$$\langle \text{div } q, v \rangle := -\langle q, \nabla v \rangle \quad \forall v \in \mathcal{C}_{0,A}^\infty(D),$$

lies in $H_A^{-1}(D)$.

Proof. As we will see in Theorem 3.18, for $v \in H_{0,A}(\text{curl}; D)$ there exists a *regular decomposition*

$$v = \nabla \phi + z, \quad \|\phi\|_{H^1(D)} + \|z\|_{H^1(D)} \simeq \|v\|_{H(\text{curl}; D)}$$

for some $\phi \in H^1_{0,A}(D)$ and $z \in [H^1_{0,A}(D)]^d$. We analyze the dual norm of $H_{0,A}(\mathrm{curl}; D)$. For better readability, we drop the subscript in the duality product $\langle .,. \rangle_{H_{0,A}(\mathrm{curl};D)}$. Let $q \in [H_{0,A}(\mathrm{curl};D)]^*$, then we obtain for its dual norm

$$\begin{aligned}
\|q\|_{[H_{0,A}(\mathrm{curl};D)]^*} &= \sup_{v \in H_{0,A}(\mathrm{curl};D)} \frac{\langle q, v \rangle}{\|v\|_{H(\mathrm{curl};D)}} \\
&\simeq \sup_{\substack{\phi \in H^1_{0,A}(D) \\ z \in [H^1_{0,A}(D)]^d}} \frac{\langle q, \nabla\phi + z \rangle}{\|\phi\|_{H^1(D)} + \|z\|_{H^1(D)}} \\
&\simeq \sup_{\phi \in H^1_{0,A}(D)} \frac{\langle q, \nabla\phi \rangle}{\|\phi\|_{H^1(D)}} + \sup_{z \in [H^1_{0,A}(D)]^d} \frac{\langle q, z \rangle}{\|z\|_{H^1(D)}} \\
&= \sup_{\phi \in H^1_{0,A}(D)} \frac{\langle \mathrm{div}\, q, \phi \rangle}{\|\phi\|_{H^1(D)}} + \sup_{z \in [H^1_{0,A}(D)]^d} \frac{\langle q, z \rangle}{\|z\|_{H^1(D)}} \\
&= \|\mathrm{div}\, q\|_{H^{-1}_A(D)} + \|q\|_{H^{-1}_A(D)}.
\end{aligned}$$

$\square$

3.1.6 Inequalities

We list some important inequalities on Sobolev spaces, such as the Poincaré and Friedrichs inequalities on $H^k(D)$, and Korn's inequality on $[H^1(D)]^d$. We will heavily use these inequalities for our analysis, both in the continuous and in the finite element setting. We state them on a domain D of unit size, and take care of proper scalings when needed. Poincaré's and Friedrichs' inequalities can be found in different forms e. g. in [Gri85, GR86, TW05].

We first state Friedrichs' inequality. It yields, that on the subspace of H^k satisfying homogenous boundary conditions for the functions at least on a part of the boundary, semi-norm and norm are equivalent.

Theorem 3.14 (Friedrichs' inequality). *Let $D \subset \mathbb{R}^d, d > 0$ be a connected, bounded Lipschitz domain of unit size $\mathrm{diam}(D) = 1$. Let $A \subset \partial D$ be a non-trivial part of the boundary. Then, for $k \in \mathbb{N}$, there exists a constant $c_F > 0$ depending only on the shape of D, the boundary part A, and k such that*

$$\|w\|_{H^k(D)} \leq c_F |w|_{H^k(D)} \qquad \forall w \in H^k_0(D). \tag{3.14}$$

Poincaré's inequality can be used on spaces where no boundary conditions are prescribed.

Theorem 3.15 (Poincaré's inequality). *Let $D \subset \mathbb{R}^d, d > 0$ be a connected, bounded Lipschitz domain of unit size $\mathrm{diam}(D) = 1$. Then there exists a constant $c_p > 0$ depending only on the shape of D, such that*

$$\|w\|_{H^1(D)} \leq c_P \left(|w|^2_{H^1(D)} + \left(\int_D w\, dx \right)^2 \right)^{1/2} \qquad \forall w \in H^1(D). \tag{3.15}$$

Korn's inequality states, that the gradient of a vector-valued function is bounded from above by the strain tensor. For a proof of this theorem, we refer to [Nit81, DL76].

Theorem 3.16 (Korn's inequality). *Let $D \subset \mathbb{R}^d, d = 2, 3$ be a connected, bounded Lipschitz domain of unit size $\operatorname{diam}(D) = 1$. Then there exists a constant $c_K > 0$ depending only on the shape of D, such that*

$$\|\varepsilon(v)\|_{L^2(D)}^2 + \|v\|_{L^2(D)}^2 \geq c_K^2 \|v\|_{H^1(D)}^2 \qquad \forall v \in [H^1(D)]^d. \tag{3.16}$$

The constant of boundedness depends on the shape of the domain: as the aspect ratio deteriorates, so does the Korn constant.

Remark 3.17. *Assuming that $A \subset \partial D$ is a nontrivial part of the boundary, one can deduce*

$$\|\varepsilon(v)\|_{L^2(D)}^2 \geq \tilde{c}_K^2 |v|_{H^1(D)}^2 \qquad \forall v \in [H^1_{0,A}(D)]^d. \tag{3.17}$$

The proof of this and similar inequalities is usually done using the compact embedding of $H^1(\Omega)$ in $L^2(\Omega)$. It moreover uses Friedrichs' and Korn's inequalities.

3.1.7 Regular decompositions of vector fields

In this section, we provide a regular decomposition of vector fields satisfying mixed boundary conditions. Such decompositions satisfying homogenous Dirichlet or Neumann boundary conditions have been shown e.g. by [PZ02, Hip02]. We utilize an extension operator first introduced in [Sch08] to obtain results also in case of mixed boundary conditions.

Theorem 3.18 (regular decomposition). *Let $D \subset \mathbb{R}^d$ satisfy Assumption 3.2, and let $A \subset \partial D$ be a boundary part. For $u \in H_{0,A}(\operatorname{curl}; D)$ there exists a decomposition*

$$u = \nabla \varphi + z,$$

where $\varphi \in H^1_{0,A}(D)$ and $z \in [H^1_{0,A}(D)]^d$. The respective parts can be bounded by

$$\|\varphi\|_{H^1} \leq c \|u\|_{H(\operatorname{curl})} \qquad \text{and} \qquad \|z\|_{H^1} \leq c \|\operatorname{curl} u\|_{L^2(D)}.$$

Proof. A proof for this theorem with homogenous Dirichlet boundary conditions (i.e. $A = \partial D$) is given in [PZ02], Neumann boundary conditions are treated in [Hip02].

We now proceed to the case of mixed boundary conditions. One can deduce the statement of the theorem for non-trivial $A \subset \partial D$ by extending D to some domain $D' \supset D$, see Figure 3.1. Here, $D_A \subset D'$ is an outer neighborhood of the Dirichlet boundary A, while $\tilde{D} = D' \backslash (D \cup D_A)$ is a stripe around domain and neighborhood. In [Sch08], an $H(\operatorname{curl})$ conforming bounded extension operator was constructed, which allows to extend u to $u' \in H(\operatorname{curl}; D')$ such that $u = 0$ on D_A.

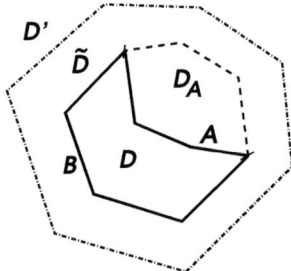

Figure 3.1: Extension of the domain D including neighborhood D_A of Dirichlet boundary A.

We now apply the theory of [Hip02] to u', and find a decomposition

$$u' = \nabla \varphi' + z',$$

where $\varphi' \in H^1(D')$ and $z' \in [H^1(D')]^d$. Note that we can avoid the usage of cohomology spaces for domains which are not simply connected, since we do not require the gauging condition $\text{div } \varphi' = 0$. Since $u' = 0$ on D_A, Lemma 2.6 in [Hip02] even ensures that $\varphi'|_{D_A} \in H^2(D_A)$. We apply the extension operator from [Ste70, Theorem 5], to obtain an extension $\tilde{\varphi}$ such that

$$\tilde{\varphi} \in H^2(D') \quad \text{and} \quad \tilde{\varphi}|_{D_A} = \varphi'|_{D_A}.$$

This allows for the decomposition

$$u = \nabla(\varphi' - \tilde{\varphi}) + z' + \nabla \tilde{\varphi}.$$

We verify that $\varphi := \varphi' - \tilde{\varphi}$ and $z := z' + \nabla \tilde{\varphi}$ satisfy homogenous boundary conditions on A. Since $\varphi = \varphi' - \tilde{\varphi} = 0$ on the neighborhood D_A of A, this implies $\varphi|_D \in H^1_{0,A}(D)$. For $z := z' + \nabla \tilde{\varphi}$, we observe that

$$z = z' + \nabla \tilde{\varphi} = -\nabla(\varphi' - \tilde{\varphi}) = 0 \quad \text{on } D_A,$$

which implies $z|_D \in [H^1_{0,A}(D)]^d$. □

3.2 Variational formulations – State of the art

In this section, we first recall the theorem by Lax and Milgram, which ensures existence and uniqueness for coercive variational problems. Then, we move to Brezzi's abstract theory for mixed problems. We then introduce several mixed formulations of the elasticity problem, which fit into this framework. All these formulations are well-studied: the first one is equivalent to the pure displacement formulation, whereas the second one is the Hellinger-Reissner formulation of elasticity. The third one is a method where the symmetry of the stress tensor is imposed weakly. We will shortly comment on existence and uniqueness of solutions, as well as stability matters. This analysis will motivate our choice of approximation spaces and the corresponding new variational formulation.

Let now $\Omega \subset \mathbb{R}^d, d = 2, 3$ be a connected, Lipschitzian domain satisfying Assumption 3.2. Let $\Gamma_D \subseteq \partial \Omega$ be a non-trivial part of its boundary, and let $\Gamma_N := \partial \Omega \backslash \Gamma_N$ be the remainder. For the elasticity problem, we impose boundary conditions

$$u = u_D \text{ on } \Gamma_D, \qquad \sigma_n = \vec{t}_N \text{ on } \Gamma_N.$$

For all the different formulations, we specify Hilbert spaces V, Σ for the displacements and stresses, respectively. We then define suitable subspaces V_0, Σ_0 satisfying the respective essential, homogenous boundary conditions on Γ_D, Γ_N. The solution spaces V_D, Σ_N are the manifolds where essential boundary conditions are fulfilled. Due to the respective extension and trace theorems, we will see that they are of the form

$$V_D = \tilde{u}_D + V_0, \qquad \Sigma_N = \tilde{\vec{t}}_N + \Sigma_0,$$

where $\tilde{u}_D, \tilde{\vec{t}}_N$ are extensions of $u_D, \vec{t}_N$ from the boundary into the full spaces V, Σ.

3.2.1 Existence and uniqueness for variational problems

We give a fundamental result on existence and uniqueness for coercive, linear variational equations. We consider the following problem, where $V_0 \subset V$ are Hilbert spaces, and $a : V \times V \to \mathbb{R}$ and $F : V \to \mathbb{R}$ are bilinear and linear forms on V.

Problem 3.19 (Primal variational problem). *Find $u \in V_0$ such that*

$$a(u, v) = \langle F, v \rangle_V \qquad \forall v \in V_0.$$

The following theorem is used as a basic tool in many proofs in numerical analysis. It can be found in almost every textbook dealing with numerical methods for partial differential equations, among them [BS02, Joh87, Bra92]

Theorem 3.20 (Lax-Milgram). *Let $V_0 \subset V$ be Hilbert spaces, and let $a : V \times V \to \mathbb{R}$ be bounded and coercive, i.e.*

$$\begin{aligned} a(u, v) &\leq c_{a,2} \|u\|_V \|v\|_V & \forall u, v \in V_0, \\ a(v, v) &\geq c_{a,1} \|v\|_V^2 & \forall v \in V_0. \end{aligned}$$

Let moreover $F \in V_0^*$, then there exists a unique solution $u \in V$ to Problem 3.19. It satisfies the stability bound

$$\|u\|_V \leq \frac{1}{c_{a,1}} \|F\|_{V_0^*}.$$

3.2.2 Brezzi's theory of mixed methods

In the following, we recall basic facts on mixed systems, following the abstract theory in [Bre74, BF91]. Let V_0, Σ_0 be Hilbert spaces equipped with norms $\|.\|_V, \|.\|_\Sigma$. We analyze the following problem.

Problem 3.21 (Abstract mixed formulation). *Find* $u \in V_0, \sigma \in \Sigma_0$ *such that*

$$
\begin{aligned}
a(\sigma,\tau) + b(\tau,u) &= \langle F_1,\tau\rangle_\Sigma & \forall \tau \in \Sigma_0, \\
b(\sigma,v) &= \langle F_2,v\rangle_V & \forall v \in V_0.
\end{aligned}
\quad (3.18)
$$

The bilinear forms naturally introduce linear operators $A : \Sigma_0 \to \Sigma_0^*$, $B : \Sigma_0 \to V_0^*$ via the relations

$$
\begin{aligned}
\langle A\sigma,\tau\rangle_\Sigma &= a(\sigma,\tau) & \forall \sigma,\tau \in \Sigma_0, \\
\langle B\tau,v\rangle_V &= b(\tau,v) & \forall \tau \in \Sigma_0, v \in V_0.
\end{aligned}
$$

The following assumptions shall be valid for $a(\cdot,\cdot), b(\cdot,\cdot)$ throughout this subsection, they provide the basis for the application of Brezzi's theorem below.

Assumption 3.22. *Let* $\Sigma_0 \subset \Sigma, V_0 \subset V$ *be Hilbert spaces. Let* $a : \Sigma_0 \times \Sigma_0 \to \mathbb{R}$, $b : \Sigma_0 \times V_0 \to \mathbb{R}$ *be bilinear forms satisfying*

1. **boundedness of** $a(\cdot,\cdot), b(\cdot,\cdot)$: *there exist positive constants* $c_{a,2}, c_{b,2} > 0$ *such that*

$$
\begin{aligned}
a(\sigma,\tau) &\leq c_{a,2}\|\sigma\|_\Sigma\|\tau\|_\Sigma & \forall \sigma,\tau \in \Sigma_0, & \quad (3.19)\\
b(\tau,v) &\leq c_{b,2}\|\tau\|_\Sigma\|v\|_V & \forall \tau \in \Sigma_0, v \in V_0, & \quad (3.20)
\end{aligned}
$$

2. **inf-sup stability of** $b(\cdot,\cdot)$: *there exists a positive constant* $c_{b,1} > 0$ *such that*

$$
\inf_{v \in V_0} \sup_{\tau \in \Sigma_0} \frac{b(\tau,v)}{\|\tau\|_\Sigma \|v\|_V} \geq c_{b,1}, \quad (3.21)
$$

3. **coercivity of** $a(\cdot,\cdot)$ **on the kernel of** B: *there exists a positive constant* $c_{a,1} > 0$ *such that*

$$
a(\tau,\tau) \geq c_{a,1}\|\tau\|_\Sigma^2 \quad \forall \tau \in \mathrm{Ker}\, B, \quad (3.22)
$$

where the kernel of B *is defined by*

$$
\mathrm{Ker}\, B := \{\tau \in \Sigma_0 : b(\tau,v) = 0 \; \forall v \in V_0\} \subseteq \Sigma_0.
$$

Conforming [BF91, Theorem 1.1], existence and uniqueness of a solution as well as stability can then be guaranteed.

Theorem 3.23 (Brezzi's theorem). *Let $\Sigma_0 \subset \Sigma$, $V_0 \subset V$ be Hilbert spaces, and let the bilinear forms $a : \Sigma_0 \times \Sigma_0 \to \mathbb{R}$, $b : \Sigma_0 \times V_0 \to \mathbb{R}$ satisfy Assumption 3.22. Then, for $F_1 \in \Sigma_0^*, F_2 \in V_0^*$, there exists a unique solution $(\sigma, u) \in \Sigma_0 \times V_0$ to Problem 3.21. It satisfies the stability bounds*

$$\|\sigma\|_\Sigma \leq \frac{1}{c_{a,1}}\|F_1\|_{\Sigma_0^*} + \frac{1}{c_{b,1}}\left(1 + \frac{c_{a,2}}{c_{a,1}}\right)\|F_2\|_{V_0^*}, \tag{3.23}$$

$$\|u\|_V \leq \frac{1}{c_{b,1}}\left(1 + \frac{c_{a,2}}{c_{a,1}}\right)\|F_1\|_{\Sigma_0^*} + \frac{c_{a,2}}{c_{b,1}^2}\left(1 + \frac{c_{a,2}}{c_{a,1}}\right)\|F_2\|_{V_0^*}. \tag{3.24}$$

Note that coercivity of $a(\cdot, \cdot)$ is not even necessary, and can be reduced to an inf-sup condition on $\operatorname{Ker} B$. However, the above assumptions will be sufficient for our further analysis. We will embed all the following methods in this framework, and use Theorem 3.23 for the analysis.

Remark 3.24. *If the solution is not sought to lie in the Hilbert space $\Sigma_0 \times V_0$, but in a manifold $\Sigma_D \times V_D = (\tilde{t}_N, \tilde{u}_D) + \Sigma_0 \times V_0$, we apply a standard homogenization technique. We then search for $(\sigma_0, u_0) = (\sigma - \tilde{t}_N, u - \tilde{u}_D) \in \Sigma_0 \times V_0$ such that*

$$\begin{aligned} a(\sigma_0, \tau) + b(\tau, u_0) &= \langle F_1, \tau \rangle_\Sigma - a(\tilde{t}_N, \tau) - b(\tau, \tilde{u}_D) & \forall \tau \in \Sigma_0, \\ b(\sigma_0, v) &= \langle F_2, v \rangle_V - b(\tilde{t}_N, v) & \forall v \in V_0. \end{aligned} \tag{3.25}$$

The right hand sides of system (3.25) still lie in the respective dual spaces, the stability estimates (3.23), (3.24) have to be modified replacing

$$\begin{aligned} \|F_1\|_{V_0^*} & \quad by \quad & \|F_1\|_{V_0^*} + c_{a,2}\|\tilde{t}_N\|_{\Sigma_0} + c_{b,2}\|\tilde{u}_D\|_{V_0}, \\ \|F_2\|_{\Sigma_0^*} & \quad by \quad & \|F_2\|_{\Sigma_0^*} + c_{b,2}\|\tilde{t}_N\|_{\Sigma_0}. \end{aligned}$$

3.2.3 The pure displacement formulation

We recall the probably most straightforward weak formulation of the elasticity problem. There, one chooses $V_{pr} := [H^1(\Omega)]^d$ as space for the displacements, and $\Sigma_{pr} := L^2_{SYM}(\Omega)$ for the stresses. Suitable subspaces satisfying essential boundary conditions are

$$\begin{aligned} V_{pr,0} &:= [H^1_{\Gamma_D,0}(\Omega)]^d, \\ V_{pr,D} &:= \{v \in [H^1(\Omega)]^d : v|_{\Gamma_D} = u_D\}, \\ \Sigma_{pr,0} &:= \Sigma_{pr,N} := \Sigma_{pr}. \end{aligned}$$

In order to obtain the primal mixed formulation, one multiplies the first equation in the elasticity problem (3.1) by test functions $\tau \in \Sigma_{pr,0}$, and the second line (3.2) by $v \in V_{pr,0}$. Then one applies integration by parts using formula (3.12) in the second line, followed by a substitution of the known normal trace $\sigma_n = \vec{t}_N$ on Γ_N. Using that $v = 0$ on Γ_D by definition, we end up with the primal mixed formulation of elasticity.

Problem 3.25 (Primal mixed formulation of elasticity). *Find $u \in V_{pr,D}$, $\sigma \in \Sigma_{pr,N}$ such that*

$$\int_\Omega (\bar{A}\sigma) : \tau \, dx - \int_\Omega \varepsilon(u) : \tau \, dx = 0 \qquad \forall \tau \in \Sigma_{pr,0}, \qquad (3.26)$$

$$- \int_\Omega \varepsilon(v) : \sigma \, dx = - \int_\Omega f \cdot v \, dx - \int_{\Gamma_N} \vec{t}_N \cdot v \, ds \quad \forall v \in V_{pr,0}. \qquad (3.27)$$

Problem 3.25 can be embedded straightforward in the abstract theory of Subsection 3.2.2. We define for $\sigma, \tau \in \Sigma_0, v \in V_0$

$$a_{pr}(\sigma, \tau) := \int_\Omega (\bar{A}\sigma) : \tau \, dx,$$

$$b_{pr}(\tau, v) := -\int_\Omega \tau : \varepsilon(v) \, dx,$$

$$\langle F_{1,pr}, \tau \rangle_\Sigma := 0,$$

$$\langle F_{2,pr}, v \rangle_V := -\int_\Omega f \cdot v \, dx - \int_{\Gamma_N} \vec{t}_N \cdot v \, ds.$$

Now, existence and uniqueness of a solution $(\sigma, u) \in \Sigma_{pr,N} \times V_{pr,D}$ as well as stability can be obtained ensuring the conditions of Brezzi's theorem. The right hand sides clearly lie in the respective dual spaces if $f \in H^{-1}_{\Gamma_D}(\Omega)$, and $\vec{t}_N \in H^{-1/2}_{00}(\Gamma_N)$. This is ensured by the trace theorem for H^1. Also boundedness of $a(\cdot, \cdot), b(\cdot, \cdot)$ follow trivially, applying the Cauchy-Schwarz inequality.

To prove inf-sup stability for $b(\cdot, \cdot)$, Korn's and Friedrichs inequalities (3.17), (3.14) are needed: For $v \in V_0$, choose $\tau = \varepsilon(v)$, then

$$b(\tau, v) \geq \|\varepsilon(v)\|^2_\Omega \geq \tilde{c}_K \|\nabla v\|_\Omega \|\tau\|_\Omega \geq \tilde{c}_K c_F^{-1} \|v\|_V \|\tau\|_\Sigma.$$

As $\tilde{c}_K$ deteriorates with the aspect ratio of the domain Ω, we expect stability problems for anisotropic domains such as plates or shells. Finite element discretizations of the mixed primal method usually show shear locking phenomena.

Coercivity of $a(\cdot, \cdot)$ can be shown straightforward, we obtain

$$a(\tau, \tau) \geq \lambda_{min}(\bar{A}) \|\tau\|^2_\Sigma.$$

There $\lambda_{min}(\bar{A})$ is the minimal eigenvalue of the compliance tensor $\bar{A}$, which tends to zero for nearly incompressible media. As a result, we expect volume locking.

Due to the fact that $\varepsilon(u) \in L^2_{SYM}(\Omega)$ for $u \in [H^1(\Omega)]^d$ and the invertibility of $\bar{A}$, we may conclude from the first equation that

$$\sigma = \bar{A}^{-1} \varepsilon(u) \qquad \text{in } L^2_{SYM}(\Omega).$$

Thereby, we can eliminate σ from the system above and end up with the (equivalent) *pure displacement formulation* of elasticity:

Problem 3.26 (Pure displacement formulation of elasticity). *Find $u \in V_{pr,D}$ such that*

$$\int_\Omega (\bar{A}^{-1}\varepsilon(u)) : \varepsilon(v)\, dx = \int_\Omega f \cdot v\, dx + \int_{\Gamma_N} \vec{t}_N \cdot v\, ds \quad \forall v \in V_{pr,0}. \tag{3.28}$$

This problem can easily be discretized using standard continuous finite elements. But, equivalently to the primal mixed problem, we expect a deterioration of stability when treating anisotropic domains or nearly incompressible media.

3.2.4 The Hellinger-Reissner formulation of elasticity

In the pure displacement formulation, all derivatives are applied to the displacements in form of the strain tensor. It is therefore necessary, to have the displacements in $H^1(\Omega)$, the stresses are left totally discontinuous. In the Hellinger-Reissner formulation, one chooses differently: Differentiation is done with respect to the stresses, the stress space is $\Sigma = H_{SYM}(\text{div})$. The matching displacement space is then $V = [L^2(\Omega)]^d$. The boundary conditions imply to use

$$V_{HR,0} := V_{HR,D} := [L^2(\Omega)]^d,$$
$$\Sigma_{HR,0} := H_{SYM,\Gamma_N,0}(\text{div}) := \{\tau \in H_{SYM}(\text{div}) : \tau_n|_{\Gamma_N} = 0\,\},$$
$$\Sigma_{HR,N} := \{\tau \in H_{SYM}(\text{div}) : \tau_n|_{\Gamma_N} = \vec{t}_N\,\}.$$

As in Section 3.2.3, we multiply the equations of elasticity (3.1),(3.2) by test functions $v \in V_{HR,0}$ and $\tau \in \Sigma_{HR,0}$. Now, we apply Green's formula (3.12) to the first line, and substitute $u = u_D$ on the Dirichlet boundary Γ_D. As $\tau_n = 0$ on Γ_N, the following problem is obtained.

Problem 3.27 (Hellinger-Reissner formulation of elasticity). *Find $u \in V_{HR,D}, \sigma \in \Sigma_{HR,N}$ such that*

$$\int_\Omega (\bar{A}\sigma) : \tau\, dx + \int_\Omega u \cdot \text{div}\, \tau\, dx = \int_{\Gamma_D} u_D \cdot \tau_n\, ds \quad \forall \tau \in \Sigma_{HR,0}, \tag{3.29}$$

$$\int_\Omega v \cdot \text{div}\, \sigma\, dx = -\int_\Omega f \cdot v\, dx \quad \forall v \in V_{HR,0}. \tag{3.30}$$

We shortly analyze Problem 3.27 under the aspects of Theorem 3.23. Therefore, we define

$$a_{HR}(\sigma,\tau) = \int_\Omega (\bar{A}\sigma) : \tau\, dx \quad \text{for } \sigma,\tau \in \Sigma_{HR},$$

$$b_{HR}(\tau,v) = \int_\Omega \text{div}\, \tau \cdot v\, dx \quad \text{for } \tau \in \Sigma_{HR}, v \in V_{HR}.$$

Due to trace theorems, the right hand sides of equations (3.29), (3.30) are clearly in $\Sigma^*_{HR,0}, V^*_{HR,0}$, and the essential boundary conditions are suitable for the respective spaces. Proving inf-sup stability for $b_{HR}(\cdot,\cdot)$ is rather simple. A proof can be found e.g. in [Bra92].

Coercivity of $a_{HR}(\cdot,\cdot)$ is also more involved. The bilinear form is clearly not uniformly coercive on the whole space $\Sigma_{HR,0}$, as the Lamé parameter $\bar{\lambda}$ tends to infinity. However, one can prove that, on Ker B_{HR}, the constant of coercivity is independent of this quantity. Therefore, the formulation is suitable for the analysis of nearly incompressible elasticity. For a more detailed discussion of the kernel-coercivity, see [ADG84] or [BF91, IV.3]. When constructing locking free finite elements, coercivity of $a_{HR}(\cdot,\cdot)$ must also hold on the discrete kernel. As shown in [AC05, AW02, AAW08], conforming finite elements satisfying this restriction can only be provided at the cost of high polynomial order.

3.2.5 Mixed methods with weak symmetry

A major problem when constructing finite elements for the Hellinger-Reissner formulation of elasticity is the symmetry of the stress tensor σ. In order to avoid this, several authors have introduced methods where the symmetry of the stress tensor is enforced by means of Lagrangian multipliers, among them [Ste86, Ste88, ABD84, AFW07].

The main idea is to introduce a skew-symmetric matrix ω, which corresponds to the skew part of ∇u. It acts as a Lagrangian multiplier ensuring the condition "σ is symmetric". Then the equations of elasticity can be reformulated to find $u : \Omega \to \mathbb{R}^d, \sigma : \Omega \to \mathbb{R}^{d\times d}, \omega : \Omega \to \mathbb{R}_{SKW}$ such that

$$\left.\begin{array}{rl}\bar{A}\sigma - \nabla u + \omega &= 0,\\ \sigma - \sigma^T &= 0,\\ -\operatorname{div}\sigma &= f\end{array}\right\} \quad \text{in } \Omega.$$

We choose the solution spaces

$$\begin{aligned}Q_{ws} &:= L^2_{SKW}(\Omega) := \{\gamma \in [L^2(\Omega)]^{d\times d} : \gamma = -\gamma^T\},\\ V_{ws,0} &:= V_{ws,D} := [L^2(\Omega)]^d \times Q_{ws},\\ \Sigma_{ws,0} &:= [H_{\Gamma_N,0}(\operatorname{div})]^d,\\ \Sigma_{ws,N} &:= \{\tau \in [H(\operatorname{div})]^d : \tau_n|_{\Gamma_N} = \vec{t}_N\}.\end{aligned}$$

Now, similarly to the Hellinger-Reissner case, we derive a variational formulation, ending up with the following problem.

Problem 3.28 (Mixed elasticity with weak symmetry). *Find* $(\sigma,(u,\omega)) \in \Sigma_{ws,N} \times V_{ws,D}$ *such that*

$$\int_\Omega (\bar{A}\sigma) : \tau\, dx + \int_\Omega u \cdot \operatorname{div}\tau\, dx + \int_\Omega \omega : \tau\, dx = \int_{\Gamma_D} u_D \cdot \sigma_N\, ds \quad \forall \tau \in \Sigma_{ws,0}, \tag{3.31}$$

$$\int_\Omega v \cdot \operatorname{div}\sigma\, dx + \int_\Omega \gamma : \sigma\, dx = -\int_\Omega f \cdot v\, dx \quad \forall (v,\gamma) \in V_{ws,0}. \tag{3.32}$$

This problem is equivalent to the mixed problem (Problem 3.27) in the sense that for each solution (σ, u, ω) to Problem 3.28, (σ, u) is a solution ot Problem 3.27. The analysis of this problem can again be done in the scope of Theorem 3.23. Using the results for the Hellinger-Reissner formulation (Problem 3.27), solvability and stability estimates can be shown straightforward. Although the infinite-dimensional problem of weak symmetry is equivalent to the Hellinger-Reissner formulation, it allows for different discretization techniques. There, the symmetry of the stress solution is usually not guaranteed. Comparably simple finite elements with respect to this formulations are available, the probably best known one is the PEERS element [ABD84].

3.3 A new weak formulation of elasticity

In this section, we present our new formulation of the equations of elasticity. We motivate the choice of spaces, and provide some basic results on their properties. We show existence and uniqueness of a solution as well as stability estimates.

Our new method will be situated "in between" the primal mixed method, where all smoothness assumptions are put on the displacement field, and the Hellinger-Reissner formulation, where the setting is vice versa, and the stresses have to satisfy a differentiability property. In both cases, the bilinear form $a(\cdot,\cdot)$ is defined in the same way, namely as L^2 product between stress and strain. When considering homogeneous, isotropic, linear elastic materials, the compliance tensor $\bar{A}$ is determined by the Lamé constants $\bar{\mu}, \bar{\lambda}$, which yields

$$a(\sigma, \tau) := \int_\Omega (\bar{A}\sigma) : \tau \, dx = \int_\Omega \frac{1}{2\bar{\mu}} \operatorname{dev} \sigma : \operatorname{dev} \tau \, dx + \int_\Omega \frac{1}{d\bar{\lambda} + 2\bar{\mu}} \operatorname{tr}(\sigma) \operatorname{tr}(\tau) \, dx. \tag{3.33}$$

The bilinear form $b(\cdot, \cdot)$ can be viewed as a duality product between the divergence of the stress and the displacement,

$$b(\tau, v) = \langle \operatorname{div} \tau, v \rangle_V.$$

For the primal mixed method, the duality product reads

$$b(\tau, v) = \langle \operatorname{div} \tau, v \rangle_{H^1(\Omega)} = -\int_\Omega \tau : \varepsilon(v) \, dx,$$

whereas in the Hellinger-Reissner setting it reads

$$b(\tau, v) = \langle \operatorname{div} \tau, v \rangle_{L^2(\Omega)} = \int_\Omega \operatorname{div} \tau \cdot v \, dx.$$

We choose now the displacement space as

$$V := H(\operatorname{curl}; \Omega),$$

with subspace $V_0 := H_{0,\Gamma_D}(\operatorname{curl})$ and manifold $V_D := \{v \in H(\operatorname{curl}) : v_\tau|_{\Gamma_D} = u_{D,\tau}\}$ containing the solution u. Thus, the stress space Σ has to satisfy the following properties, such that a mixed system of the form (3.18) is well defined:

1. $\Sigma \subset L^2_{SYM}(\Omega)$, such that the volume integral $(\bar{A}\sigma, \tau)_\Omega$ can be evaluated

2. For $\tau \in \Sigma$, we need that $\operatorname{div} \tau \in V_0^* = H^{-1}_{\Gamma_D}(\operatorname{div})$; this ensures that the duality product $\langle \operatorname{div} \tau, v \rangle_V$ is well defined.

We introduce a new space, which consists of functions satisfying these two conditions.

Definition 3.29. *Let $D \subset \mathbb{R}^d$ be a domain satisfying Assumption 3.2. We define the Hilbert space*

$$H(\operatorname{div} \operatorname{div}; D) := \{\tau \in L^2_{SYM}(D) : \operatorname{div} \operatorname{div} \tau \in H^{-1}(D)\}. \tag{3.34}$$

The space is equipped with norm

$$\begin{aligned}
\|\tau\|^2_{H(\operatorname{div} \operatorname{div}; D)} &:= \|\tau\|^2_{L^2(D)} + \|\operatorname{div} \operatorname{div} \tau\|^2_{H^{-1}(\operatorname{div}; D)} \\
&= \|\tau\|^2_{L^2(D)} + \sup_{v \in H_0(\operatorname{curl}; D)} \frac{\langle \operatorname{div} \tau, v \rangle^2_{H(\operatorname{curl}; D)}}{\|v\|^2_{H(\operatorname{curl}; D)}}.
\end{aligned}$$

In this definition, the divergence of the divergence of a tensor-valued function is a scalar field. For the case of a symmetric tensor in two space dimensions, it reads

$$\operatorname{div} \operatorname{div} \tau = \operatorname{div} \begin{pmatrix} \frac{\partial \tau_{11}}{\partial x_1} + \frac{\partial \tau_{12}}{\partial x_2} \\ \frac{\partial \tau_{12}}{\partial x_1} + \frac{\partial \tau_{22}}{\partial x_2} \end{pmatrix} = \frac{\partial^2 \tau_{11}}{\partial x_1^2} + 2 \frac{\partial^2 \tau_{12}}{\partial x_1 \partial x_2} + \frac{\partial^2 \tau_{22}}{\partial x_2^2}.$$

We search for the stresses in the space

$$\Sigma := H(\operatorname{div} \operatorname{div}; \Omega).$$

As we will see later, it is possible to define a normal-normal trace operator tr_{nn} on the space $H(\operatorname{div} \operatorname{div}; \Omega)$. Recall the normal-normal and normal-tangential trace of some smooth, tensor-valued field τ,

$$\operatorname{tr}_{nn}(\tau) := \tau_{nn} := \tau_n n = n^T \tau n, \qquad \operatorname{tr}_{n\tau}(\tau) := \tau_{n\tau} := \tau_n - \tau_{nn} n.$$

Analogously, we will provide an inverse trace theorem, which allows to lift any distribution from a suitable trace space to the volume space $H(\operatorname{div} \operatorname{div})$. Therefore, we define the spaces

$$\begin{aligned}
\Sigma_0 &:= H_{0,\Gamma_N}(\operatorname{div} \operatorname{div}; \Omega) := \{\tau \in L^2_{SYM}(\Omega) : \operatorname{div} \operatorname{div} \tau \in H^{-1}_{\Gamma_D}(\Omega), \tau_{nn}|_{\Gamma_N} = 0\}, \\
\Sigma_N &:= \{\tau \in L^2_{SYM}(\Omega) : \operatorname{div} \operatorname{div} \tau \in H^{-1}_{\Gamma_D}(\Omega), \tau_{nn}|_{\Gamma_N} = \vec{t}_{N,n}\}.
\end{aligned}$$

We note that

$$\|\tau\|^2_{H(\text{div div};\Omega),\Gamma_N} := \|\tau\|^2_{L^2(\Omega)} + \|\operatorname{div}\tau\|^2_{H^{-1}_{\Gamma_D}(\text{div};\Omega)}$$

$$= \|\tau\|^2_{L^2(\Omega)} + \sup_{v\in H_{0,\Gamma_D}(\text{curl};\Omega)} \frac{\langle\operatorname{div}\tau,v\rangle^2_{H(\text{curl};\Omega)}}{\|v\|^2_{H(\text{curl};\Omega)}}$$

is a suitable norm for the space $H_{0,\Gamma_N}(\text{div div};\Omega)$.

We use the spaces $V = H(\text{curl})$ and $\Sigma = H(\text{div div})$ and their respective subspaces V_0, V_D and Σ_0, Σ_N to derive a new variational formulation. We have already seen in Theorem 3.12, that the choice of $H(\text{curl})$ means we have tangential continuity for the displacements. We will see that for the stresses, the choice of spaces implies normal-normal continuity, i.e. we have σ_{nn} continuous across interfaces. Therefore, we speak of the Tangential-Displacement-Normal-Normal-Stress (TD-NNS) formulation of elasticity.

Problem 3.30 (TD-NNS formulation of elasticity). *Find $\sigma \in \Sigma_N$, $u \in V_D$ such that*

$$\int_\Omega (\bar{A}\sigma):\tau\,dx + \langle\operatorname{div}\tau, u\rangle_{V_0} = \int_{\Gamma_D} u_{D,n}\,\tau_{nn}\,ds \qquad \forall \tau \in \Sigma_0, \qquad (3.35)$$

$$\langle\operatorname{div}\sigma, v\rangle_{V_0} = -\int_\Omega f\cdot v\,dx + \int_{\Gamma_N} \vec{t}_{N,\tau}\cdot v_\tau \qquad \forall v \in V_0. \qquad (3.36)$$

This formulation again fits into the framework of Section 3.2.2. It is a problem of form (3.18), where for $\sigma, \tau \in \Sigma_0$, $v \in V_0$

$$a(\sigma,\tau) := \int_\Omega (\bar{A}\sigma):\tau\,dx, \qquad (3.37)$$

$$b(\tau,v) := \langle\operatorname{div}\sigma, v\rangle_{H(\text{curl})}, \qquad (3.38)$$

$$F_1(\tau) := \int_{\Gamma_D} u_{D,n}\,\tau_{nn}\,ds, \qquad (3.39)$$

$$F_2(v) := -\int_\Omega f\cdot v\,dx + \int_{\Gamma_N} \vec{t}_{N,\tau}\cdot v_\tau. \qquad (3.40)$$

In order to show existence and uniqueness of a solution, as well as stability estimates, we need to investigate for properties of the space $H(\text{div div})$.

3.3.1 The space $H(\text{div div})$

Throughout the following, let $D \subset \mathbb{R}^d$, $d = 2, 3$ be a connected, Lipschitzian domain of unit size, diam $D \simeq 1$. Let D satisfy Assumption 3.2, then the boundary ∂D is either smooth or polyhedral. Let $A \subseteq \partial D$ denote a closed part of its boundary, while $B = \partial D \setminus A$ denotes the remainder. First, we are concerned with the evaluation of the duality product $\langle\operatorname{div}\tau, v\rangle_{H(\text{curl};D)}$ for smooth functions. We see that it can be done by means of surface and volume integrals. We provide a trace and an inverse trace theorem for the space $H(\operatorname{div}\operatorname{div};D)$, as well as an integration by parts formula. From these results we derive interface conditions for piecewise defined spaces.

Lemma 3.31 (Evaluation of the duality product). *For some $\tau \in H_{0,B}(\operatorname{div}\operatorname{div};D)$ and $v \in H_{0,A}(\operatorname{curl};D)$ smooth enough, the duality product $\langle \operatorname{div}\tau, v \rangle_{H(\operatorname{curl};D)}$ can be evaluated as follows*

$$\langle \operatorname{div}\tau, v\rangle_{H(\operatorname{curl};D)} = \int_D \operatorname{div}\tau \cdot v\, dx - \int_B \tau_{n\tau} v_\tau\, ds \quad (3.41)$$

$$= -\int_D \tau : \varepsilon(v)\, dx + \int_A \tau_{nn} v_n\, ds. \quad (3.42)$$

Remark 3.32. *The first evaluation (3.41) is still true for $v \in H_{0,A}(\operatorname{curl};D)$ and tensor fields $\tau \in H_{SYM,0,B}(\operatorname{div};D)$, if $\tau_{n\tau} \in H_\|^{-1/2}(\operatorname{div}_B; B)$ and the surface integrals are understood as duality pairings. As we will see in Theorem 3.34, the normal-normal trace of $\tau \in H(\operatorname{div}\operatorname{div}; D)$ lies in the dual space of the Neumann trace space of $H^2(D) \cap H_{0,A}^1(D)$, which is $H_{n,0,A}^{1/2}(\partial D)$ by its definition (3.7). The second form of evaluation (3.42) is then possible for $v \in [H_{0,A}^1(D)]^d$ and $\tau \in H_{0,B}(\operatorname{div}\operatorname{div};D)$. These statements follow, as the set of smooth functions lies dense in the respective spaces.*

Proof of Lemma 3.31. Let $\tau \in H_{0,B}(\operatorname{div}\operatorname{div};D), v \in H_{0,A}(\operatorname{curl};D)$ be smooth. We first show representation (3.42), the second representation follows then from integration by parts, see formula (3.12). The distributional divergence is defined such that for any function τ and $\phi \in C_{0,\tau}^\infty(\bar{D})$

$$\langle \operatorname{div}\tau, \phi\rangle_{H(\operatorname{curl};D)} = -\int_D \tau:\varepsilon(\phi)\,dx + \int_{\partial D}\tau_n \phi\,ds$$

$$= -\int_D \tau:\varepsilon(\phi)\,dx + \int_{\partial D}\tau_{nn}\phi_n\,ds + \int_{\partial D}\tau_{n\tau}\underbrace{\phi_\tau}_{=0}\,ds.$$

We take this definition for arbitrary, sufficiently smooth v, τ, which satisfy the boundary conditions $v_\tau = 0$ on A, $\tau_{nn} = 0$ on B. They obviously satisfy identities (3.41) and (3.42). $\square$

Remark 3.33. *We note that the evaluation of the duality product $\langle \operatorname{div}\tau, v\rangle_{H(\operatorname{curl};D)}$ by means of equation (3.41) for general $v \in H(\operatorname{curl})$ is only possible if the normal-tangential component $\tau_{n\tau}$ of the stress tensor lies in $H_\|^{-1/2}(\operatorname{div}_{\partial D}; \partial D)$. This condition may be violated even for smooth τ on domains with corners.*

The following trace theorem states, that functions in $H(\operatorname{div}\operatorname{div})$ have a normal-normal trace.

Theorem 3.34 (Trace theorem for $H(\operatorname{div}\operatorname{div})$). *Let $D \subset \mathbb{R}^d, d = 2,3$ be a bounded Lipschitz domain with polyhedral boundary according to Assumption 3.2, $\operatorname{diam} D \simeq 1$.*

1. *The trace operator $\operatorname{tr}_{\partial D, nn}$ is well defined the space on $H(\operatorname{div}\operatorname{div}; D)$ as an extension from $C_{SYM}^\infty(\bar{D})$. It is continuous onto $H_n^{-1/2}(\partial D)$, there exists a constant $c_{tr_{nn}} > 0$ such that*

$$\|\operatorname{tr}_{\partial D,nn}\tau\|_{H_n^{-1/2}(\partial D)} \le c_{tr_{nn}}\|\tau\|_{H(\operatorname{div}\operatorname{div};D)} \quad \forall \tau \in H(\operatorname{div}\operatorname{div};D).$$

2. For $g \in H_n^{-1/2}(\partial D)$, there exists some $\tau \in H(\text{div div}; D)$ such that

$$\text{tr}_{\partial D, nn} \tau = g \quad \text{on } \partial D \quad \text{and} \quad \|\tau\|_{H(\text{div div}; D)} \leq c_{extnn} \|g\|_{H_n^{-1/2}(\partial D)}.$$

The constants $c_{tr_{nn}}, c_{extnn}$ are positive and depend only on the shape of the domain D.

Proof. 1. Let $\tau \in C^\infty_{SYM}(\bar{D})$ be arbitrary. For simplicity of notation, we write τ_{nn} for $\text{tr}_{\partial D, nn} \tau$. By the definition of the dual norm, we have

$$\|\tau_{nn}\|_{H_n^{-1/2}(\partial D)} = \sup_{w \in H_{n,0}^{1/2}(\partial D)} \frac{\langle \tau_{nn}, w \rangle_{H_{n,0}^{1/2}(\partial D)}}{\|w\|_{H_{n,0}^{1/2}(\partial D)}}. \tag{3.43}$$

We note that, due to Friedrichs inequality (Theorem 3.14), the semi-norm $\|\nabla \cdot\|_{H^1(D)}$ forms an equivalent norm on $H^2(D) \cap H_0^1(D)$. From the definition of the Neumann trace space $H_{n,0}^{1/2}(\partial D)$, we have that

$$\|\tau_{nn}\|_{H_n^{-1/2}(\partial D)} = \sup_{\varphi \in H^2(D) \cap H_0^1(D)} \frac{\langle \tau_{nn}, \partial \varphi / \partial n \rangle_{H_{n,0}^{1/2}(\partial D)}}{\|\nabla \varphi\|_{H^1(D)}}.$$

We may proceed, using Green's formula, and setting $\psi = \nabla \varphi$ in the second line,

$$\begin{aligned}
\|\tau_{nn}\|_{H_n^{-1/2}(\partial D)} &= \sup_{\varphi \in H^2(D) \cap H_0^1(D)} \frac{(\text{div } \tau, \nabla \varphi)_D + (\tau, \varepsilon(\nabla \varphi))_D - \langle \tau_{n\tau}, \partial \varphi / \partial \tau \rangle}{\|\nabla \varphi\|_{H^1(D)}} \\
&\leq \sup_{\psi \in [H_0^1(D)]^d} \frac{(\text{div } \tau, \psi)_D}{\|\psi\|_{H^1(D)}} + \sup_{\varphi \in H^2(D) \cap H_0^1(D)} \frac{(\tau, \varepsilon(\nabla \varphi))_D}{\|\nabla \varphi\|_{H^1(D)}} + 0 \\
&\leq \|\text{div } \tau\|_{H^{-1}(D)} + \|\tau\|_{L^2(D)} \\
&\leq c \|\tau\|_{H(\text{div div}; D)}
\end{aligned}$$

Here, we used that the tangential derivative $\partial \varphi / \partial \tau$ vanishes for $\varphi \in H_0^1(D)$. This implies that the normal-normal trace operator is continuous with respect to the norm on $H(\text{div div}; D)$, and can therefore be extended to the full space.

2. Let $g \in H_n^{-1/2}(\partial D)$, and let $\tilde{w} \in H^2(D)$ be a weak solution to

$$\begin{aligned}
\text{div div } \nabla^2 \tilde{w} + \tilde{w} &= 0 \quad \text{in } D, \\
w &= 0 \quad \text{on } \partial D, \\
(\nabla^2 \tilde{w})_{nn} &= g \quad \text{on } \partial D,
\end{aligned}$$

where $\nabla^2 \tilde{w}$ denotes the Hessian of $\tilde{w}$. The weak formulation of this problem reads

$$\int_D \nabla^2 \tilde{w} : \nabla^2 \tilde{v} \, dx + \int_D \tilde{w} \, \tilde{v} \, dx = \int_{\partial D} g \frac{\partial \tilde{v}}{\partial n} \, ds \quad \forall \tilde{v} \in H^2(D) \cap H_0^1(D). \tag{3.44}$$

It is well posed, the bilinear form coincides with the H^2 inner product $(\cdot,\cdot)_{H^2(D)}$, and g lies in the dual of the Neumann trace space of $H^2(D) \cap H_0^1(D)$. As $\tilde{w} \in H^2(D)$, we have $\partial \tilde{w}/\partial n \in H_{n,0}^{1/2}(\partial D)$. As $\tilde{w}$ vanishes on the boundary, we may estimate, using the definition of the norm (3.6) for the Neumann trace space $H_{n,0}^{1/2}(\partial D)$

$$\|\tilde{w}\|_{H^2(D)} \geq \left\|\frac{\partial \tilde{w}}{\partial n}\right\|_{H_n^{-1/2}(\partial D)}.$$

By the Lemma of Lax and Milgram (Theorem 3.20), we conclude

$$\|\tilde{w}\|_{H^2(D)} \leq c\|g\|_{H_n^{-1/2}(\partial D)}.$$

Then we have, for $\tau = \nabla^2 w$,

$$\|g\|_{H_n^{-1/2}(\partial D)} \geq c\|\tilde{w}\|_{H^2(D)} \geq c\|\tau\|_{L^2(D)}.$$

As we have $\operatorname{div}\operatorname{div}\tau = -\tilde{w}$, we obtain

$$\|g\|_{H_n^{-1/2}(\partial D)} \geq c\|\tilde{w}\|_{L^2(D)} \geq c\|\operatorname{div}\operatorname{div}\tau\|_{H^{-1}(D)},$$

which proves the second statement. $\square$

Theorem 3.35 (Integration by parts). *Let $D \subset \mathbb{R}^d, d = 2, 3$ be a bounded Lipschitz domain satisfying Assumption 3.2, $\operatorname{diam} D \simeq 1$. Let $v \in H(\operatorname{curl}; D)$ and $\tau \in H(\operatorname{div}\operatorname{div}; D)$. In case of additional smoothness, the following integration by parts formulae hold true.*

1. *For $v \in [\mathcal{C}^1(\bar{D})]^d$,*

$$\langle \operatorname{div} \tau, v \rangle_{H(\operatorname{curl})} = -\int_D \tau : \varepsilon(v)\, dx + \int_{\partial D} \tau_{nn} v_n\, ds. \tag{3.45}$$

2. *For $\tau \in \mathcal{C}^1_{SYM}(\bar{D})$ with $\tau_{n\tau} \in H_\|^{-1/2}(\operatorname{div}_{\partial D}; \partial D)$*

$$\langle \operatorname{div} \tau, v \rangle_{H(\operatorname{curl})} = \int_D \operatorname{div} \tau \cdot v\, dx - \int_{\partial D} \tau_{n\tau} \cdot v_\tau\, ds. \tag{3.46}$$

Proof. Both integration by parts formulae hold true for smooth v, τ, see Theorem 3.11. Due to trace theorems (Theorem 3.8 and Theorem 3.34), we have boundedness of the respective boundary integrals on the right hand sides of equations (3.45) and (3.46) in the sense of duality products. By density arguments, they can be extended to the full spaces $H(\operatorname{curl}), H(\operatorname{div}\operatorname{div})$. $\square$

Theorem 3.36 (Interface conditions for $H(\operatorname{div}\operatorname{div})$). *Let $\mathcal{D} = \{D_1, \ldots, D_N\}$ be a non-overlapping domain decomposition for D, i.e. $D_i \cap D_j = \emptyset$ for $i \neq j \in \{1, \ldots, N\}$ and $\bar{D} = \bigcup \bar{D}_i$. Let*

$\Gamma_{ij} := \partial D_i \cap \partial D_j$ be the common interface of any two sub-domains $D_i, D_j \in \mathcal{D}$. Let τ be defined piecewise such that $\tau_i := \tau|_{D_i} \in H(\mathrm{div\,div}; D_i)$, and let moreover $\mathrm{div}\,\tau_i$ lie in $[H(\mathrm{curl}; D_i)]^*$ for $i = 1, \ldots, N$. Then, if
$$\mathrm{tr}_{\Gamma_{ij},nn}\,\tau_i = \mathrm{tr}_{\Gamma_{ij},nn}\,\tau_j \quad \text{for } i,j = 1, \ldots N,$$
the composite function τ lies in the global space $H(\mathrm{div\,div}; D)$.

Proof. Clearly, the composite function τ lies in $L^2(D)$, as the restrictions to the respective sub-domains do. Thus, it remains to verify that $\mathrm{div\,div}\,\tau \in H^{-1}(D)$. Equivalently, we will prove
$$\mathrm{div}\,\tau \in H^{-1}(\mathrm{div}) = [H_0(\mathrm{curl})]^*.$$
Let therefore $\phi \in [\mathcal{C}_0^\infty(D)]^d$ be a smooth test function. Due to the integration by parts formula given in Theorem 3.35, we have
$$\begin{aligned}
\langle \mathrm{div}\,\tau, \phi \rangle_{H(\mathrm{curl})} &= -\int_D \tau : \varepsilon(\phi)\,dx + \int_{\partial D} \tau_{nn}\phi_n\,ds \\
&= -\sum_{i=1}^N \int_{D_i} \tau : \varepsilon(\phi)\,dx + \int_{\partial D} \tau_{nn}\phi_n\,ds + \sum_{\Gamma_{ij}} \int_{\Gamma_{ij}} \underbrace{(\tau_{n_i n_i} - \tau_{n_j n_j})}_{=0}\phi_n\,ds.
\end{aligned}$$
Here, the surface integrals are to be understood as duality products. We may reorder the surface terms sub-domain by sub-domain, and obtain, using again the integration by parts formula from Theorem 3.35
$$\begin{aligned}
\langle \mathrm{div}\,\tau, \phi \rangle_{H(\mathrm{curl})} &= \sum_{i=1}^N \left[-\int_{D_i} \tau : \varepsilon(\phi)\,dx + \int_{\partial D_i} \tau_{nn}\phi_n\,ds \right] \\
&= \sum_{i=1}^N \langle \mathrm{div}\,\tau_i, \phi \rangle_{H(\mathrm{curl}, D_i)} \leq \left[\sum_{i=1}^N \|\mathrm{div}\,\tau_i\|_{[H(\mathrm{curl}; D_i)]^*} \right] \|\phi\|_{H(\mathrm{curl}; D)}.
\end{aligned}$$
This implies $\mathrm{div}\,\tau \in [H_0(\mathrm{curl})]^*$. □

Remark 3.37. *Note that $\tau_i \in H(\mathrm{div\,div}; D_i)$ for $i = 1, \ldots, N$ does not imply the condition "$\mathrm{div}\,\tau_i \in [H(\mathrm{curl}; D_i)]^*$", but only "$\mathrm{div}\,\tau_i \in [H_0(\mathrm{curl}; D_i)]^*$". For a piecewise smooth tensor field τ, "$\mathrm{div}\,\tau_i \in [H(\mathrm{curl}; D_i)]^*$" is satisfied iff its normal-tangential trace $\tau_{i,n\tau}$ lies in $H_\|^{-1/2}(\mathrm{div}_{\partial D_i}; \partial D_i)$. Then, the duality product $\langle \mathrm{div}\,\tau_i, v_i \rangle_{H(\mathrm{curl})}$ for $v_i \in H(\mathrm{curl}; D_i)$ can be evaluated using the integration by parts formula (3.46) provided in Theorem 3.35. Note that this additional condition needs not be satisfied even for smooth functions on sub-domains with corners. In this regard, our finite element approximations presented in Chapter 4 will not be conforming for $H(\mathrm{div\,div})$. However, our analysis will rectify this lack of conformity.*

3.3.2 Stability analysis

We are concerned with stability analysis of our mixed formulation. In all our statements, we assume that the domain Ω is connected, with its boundary either smooth or polyhedral, according to Assumption 3.2. We verify the conditions of Assumption 3.22. Continuity of $a(\cdot,\cdot), b(\cdot,\cdot)$ are clear, as

$$a(\sigma,\tau) = \int_\Omega (\bar{A}\sigma):\tau\,dx \leq \frac{1}{2\bar{\mu}}\|\sigma\|_\Sigma \|\tau\|_\Sigma,$$
$$b(\tau,v) = \langle \operatorname{div}\tau,v\rangle_{H(\operatorname{curl})} \leq \|\tau\|_\Sigma \|v\|_V.$$

Also, an inf-sup condition for the divergence operator acting on $H(\operatorname{div}\operatorname{div})$ can be shown directly.

Lemma 3.38. *Let $V_0 = H_{0,\Gamma_D}(\operatorname{curl};\Omega), \Sigma_0 = H_{0,\Gamma_N}(\operatorname{div}\operatorname{div};\Omega)$ be the spaces satisfying homogenous essential boundary conditions on Γ_D, Γ_N, respectively. The bilinear form $b(\cdot,\cdot)$ defined by relation (3.38) is inf-sup stable on $\Sigma_0 \times V_0$, i.e. there exists a constant $c_{b,1} > 0$ such that*

$$\inf_{v\in V_0}\sup_{\tau\in\Sigma_0} \frac{b(\tau,v)}{\|\tau\|_\Sigma\|v\|_V} > c_{b,1} > 0.$$

Proof. Let $\mathcal{I}$ denote the Riesz isomorphism from V_0 to V_0^*. It is defined by the relation

$$\langle q,v\rangle_V = (q,\mathcal{I}v)_{V_0^*} \qquad \forall q \in V_0^*.$$

Being an isomorphism, it moreover satisfies $\|\mathcal{I}v\|_{V_0^*} = \|v\|_V$ for any $v \in V_0$. For our choice of spaces, $\mathcal{I}$ maps $v \in V_0 = H_{0,\Gamma_D}(\operatorname{curl})$ to $\mathcal{I}v \in H_{\Gamma_D}^{-1}(\operatorname{div})$.

Let now $v \in V_0$ be fixed; in order to show the inf-sup condition it suffices to find some $\tau \in \Sigma_0$ such that

$$b(\tau,v) \geq c_{b,1}\|\tau\|_\Sigma\|v\|_V.$$

Now, for this v, let $q := \mathcal{I}v$ be its dual element in $H_{\Gamma_D}^{-1}(\operatorname{div})$. We choose $w \in [H_{0,\Gamma_D}^1]^d$ to be a weak solution to

$$\int_\Omega \varepsilon(w):\varepsilon(v)\,dx = \int_\Omega q\cdot v\,dx \qquad \forall v \in [H_{0,\Gamma_D}^1]^d.$$

Due to Theorem 3.20 (Lax, Milgram) and Korn's inequality, this equation defines w uniquely, and we see that

$$\|\varepsilon(w)\|_{L^2(\Omega)} \simeq \|q\|_{H^{-1}(\Omega)}.$$

We emphasize that, when estimating the full norm $\|w\|_{H^1(\Omega)}$, Korn's constant c_K comes in. However, this is not necessary in our line of proof. We will end up with an estimate independent of c_K. We choose $\tau = \varepsilon(w)$, and verify that $\tau \in \Sigma_0$:

- The tensor τ lies in $L^2_{SYM}(\Omega)$, as

$$\|\tau\|_{L^2(\Omega)} \simeq \|q\|_{H^{-1}(\Omega)} \leq \|v\|_V.$$

- Since $\operatorname{div}\tau = q \in H^{-1}_{\Gamma_D}(\operatorname{div})$, we know that $\operatorname{div}\operatorname{div}\tau \in H^{-1}_{\Gamma_D}(\Omega)$, and
$$\|\operatorname{div}\operatorname{div}\tau\|_{H^{-1}(\Omega)} = \|\operatorname{div}q\|_{H^{-1}(\Omega)} \leq \|q\|_{H^{-1}(\operatorname{div})} = \|v\|_V.$$

- The natural boundary condition $\tau_n = \varepsilon(w)n = 0$ on the Neumann boundary Γ_N then ensures that $\tau \in \Sigma_0$.

As $q = \operatorname{div}\tau$, we have $\|\operatorname{div}\tau\|_{H^{-1}(\operatorname{div};\Omega)} = \|v\|_V$ by definition. We can even bound $\|\tau\|_\Sigma$ by $\|\operatorname{div}\tau\|_{H^{-1}(\operatorname{div};\Omega)}$:

$$\begin{aligned}\|\tau\|^2_\Sigma &= \|\operatorname{div}\operatorname{div}\tau\|^2_{H^{-1}(\Omega)} + \|\tau\|^2_{L^2(\Omega)} \\ &\simeq \|\operatorname{div}\operatorname{div}\tau\|^2_{H^{-1}(\Omega)} + \|q\|^2_{H^{-1}(\Omega)} \\ &= \|\operatorname{div}\operatorname{div}\tau\|^2_{H^{-1}(\Omega)} + \|\operatorname{div}\tau\|^2_{H^{-1}(\Omega)} = \|\operatorname{div}\tau\|^2_{H^{-1}(\operatorname{div};\Omega)}.\end{aligned}$$

Thus, we conclude,

$$b(\tau, v) = \langle \operatorname{div}\tau, v\rangle_V = (\operatorname{div}\tau, \mathcal{I}v)_{H^{-1}(\operatorname{div})} = \|\operatorname{div}\tau\|^2_{H^{-1}(\operatorname{div})} \simeq \|\tau\|_\Sigma \|v\|_V.$$

$\square$

Last, we show coercivity of $a(\cdot,\cdot)$ on the kernel $\operatorname{Ker}B$. There, we lay special emphasis that the constant of coercivity does not deteriorate as the material becomes incompressible, i.e. as the Lamé parameter $\bar\lambda$ tends to infinity. We first prove a lemma, which can be found in [GR86, BS02] for pure Dirichlet or Neumann boundary conditions, i.e. $\partial\Omega = \Gamma_D$ or $\partial\Omega = \Gamma_N$.

Lemma 3.39. *Let $\Gamma_N = \partial\Omega\backslash\Gamma_D$ be a non-trivial part of the boundary. Then for all $p \in L^2(\Omega)$ there exists a function $v \in H^1(\Omega) \cap H_{0,\Gamma_N}(\operatorname{curl};\Omega)$ satisfying*

$$\begin{aligned}\operatorname{div}v &= p &\text{in }\Omega,\\ v_n &= 0 &\text{on }\Gamma_D.\end{aligned}$$

If $\Gamma_N = \emptyset$, and p satisfies the compatibility condition $\int_\Omega p\,dx = 0$, we can find $v \in H^1_0(\Omega)$ such that

$$\operatorname{div}v = p \quad \text{in }\Omega.$$

In both cases, there exists a positive constant c_{div} such that

$$\|v\|_{H^1(\Omega)} \leq c_{div}\|p\|_{L^2(\Omega)}.$$

Proof. The case of pure Dirichlet boundary conditions is covered by [GR86, BS02]. Note that the statement of the lemma is even stronger in this case, as we have $v \in H^1_0(\Omega)$, i.e. the full trace of v vanishes, not only its normal component.

We adapt this proof for the case of mixed boundary conditions. Let $p \in L^2(\Omega)$ be fixed, and let Γ_N be a non-trivial subset of $\partial\Omega$. Let $\bar{p} \in \mathbb{R}$ be the average value of p on Ω,

$$\bar{p} := \frac{1}{|\Omega|} \int_\Omega p\, dx.$$

Then $p - \bar{p}$ satisfies the compatibility condition above, and we deduce the existence of some $v_1 \in H_0^1(\Omega)$ such that

$$\operatorname{div} v_1 = p \quad \text{in } \Omega,$$
$$\|v\|_{H^1(\Omega)} \leq c_{div}\|p\|_{L^2(\Omega)}.$$

Secondly, we are concerned with finding some $v_2 \in H^1(\Omega)$, $v_{2,\tau} = 0$ on Γ_N, $v_{2,n} = 0$ on Γ_D, and $\operatorname{div} v_2 = \bar{p}$. To do so, let w be the unique weak solution to

$$-\Delta w = \bar{p} \quad \text{in } \Omega,$$
$$w = 0 \quad \text{on } \Gamma_N,$$
$$\frac{\partial w}{\partial n} = 0 \quad \text{on } \Gamma_D.$$

Existence and uniqueness follow from Theorem 3.20 (Lax, Milgram). Note that we required Γ_N to be non-trivial. As the right hand side is constant and all boundary conditions are homogenous, elliptic regularity [Gri85] ensures that $w \in H^2(\Omega)$ and moreover $\|w\|_{H^2(\Omega)} \leq c\|\bar{p}\|_\Omega$. We set $v_2 := \nabla w$, and observe

- v_2 lies in $H^1(\Omega)$,
- $\operatorname{div} v_2 = \bar{p}$,
- $v_{2,\tau} = \partial w/\partial \tau = 0$ on Γ_N as $w = 0$,
- $v_{2,n} = \partial w/\partial n = 0$ on Γ_D,
- $\|v_2\|_{H^1(\Omega)} \leq c\|\bar{p}\|_\Omega$.

This implies that $v = v_1 + v_2$ satisfies the statement of the lemma. $\square$

This enables us to prove coercivity of the problem of mixed elasticity.

Lemma 3.40. *The bilinear form $a(\cdot,\cdot)$ defined by relation (3.33) is coercive on the subspace $\operatorname{Ker} B \subset V_0$,*

$$a(\tau,\tau) \geq c_{a,1}\|\tau\|^2_{H(\operatorname{div}\operatorname{div})} \qquad \forall \tau \in \operatorname{Ker} B,$$

and the constant $c_{a,1} > 0$ is independent of the Lamé parameter $\bar{\lambda}$ tending to infinity. The kernel of B is the subspace of divergence free L^2 functions,

$$\operatorname{Ker} B = \{\tau \in L^2_{SYM}(\Omega) : \operatorname{div}\tau = 0 \text{ in } H^{-1}_{\Gamma_D}(\operatorname{div})\}. \tag{3.47}$$

Proof. The characterization (3.47) of Ker B follows directly from the definition of $b(\cdot, \cdot)$ and V_0. Note that, for $\tau \in \text{Ker } B$, also div div τ vanishes, and $\|\tau\|_\Sigma = \|\tau\|_{L^2(\Omega)}$. To prove coercivity of $a(\cdot, \cdot)$, we proceed similar to [BF91]. In case of pure Dirichlet boundary conditions, we restrict the stress space Σ_0 to

$$\widetilde{\Sigma}_0 := \Big\{ \tau \in \Sigma_0 : \int_\Omega \text{tr}(\tau) \, dx = 0 \Big\}.$$

This is possible as the solution σ to the homogenous problem satisfies this constraint, since

$$\int_\Omega \text{tr}(\tau) \, dx = \int_\Omega \text{tr}(\bar{A}^{-1}\varepsilon(u)) \, dx = \int_\Omega (d\bar{\lambda} + 2\bar{\mu}) \, \text{div } u \, dx = (d\bar{\lambda} + 2\bar{\mu}) \int_{\partial\Omega} u_n \, ds = 0$$

Due to Lemma 3.39 there exists some $v \in [H^1(\Omega)]^d$ such that $v_\tau = 0$ on Γ_N and $v_n = 0$ on Γ_D and

$$\text{div } v = \text{tr}(\tau), \quad \text{and} \quad \|v\|_{H^1(\Omega)} \leq c_{div} \|\text{tr}(\tau)\|_{L^2(\Omega)}.$$

Then, we estimate

$$\|\text{tr}(\tau)\|_{L^2(\Omega)}^2 = \int_\Omega \text{tr}(\tau) \, \text{div } v \, dx = \int_\Omega (\text{tr}(\tau) I) : \varepsilon(v) \, dx$$
$$= d \int_\Omega (\tau - \text{dev } \tau) : \varepsilon(v) \, dx.$$

Now, since τ lies in Ker B, its divergence vanishes in $H^{-1}_{\Gamma_D}(\text{div})$, this means, using the representation (4.19) for $v \in H^1(\Omega)$

$$0 = \langle \text{div } \tau, v \rangle_{H(\text{curl})} = \int_\Omega \tau : \varepsilon(v) \, dx - \int_{\Gamma_D} \tau_{nn} \underbrace{v_n}_{=0} ds$$

We insert this above, and obtain

$$\|\text{tr}(\tau)\|_{L^2(\Omega)}^2 = d \int_\Omega (\tau - \text{dev } \tau) : \varepsilon(v) \, dx = -d \int_\Omega \text{dev } \tau : \varepsilon(v) \, dx$$
$$\leq d \|\text{dev } \tau\|_{L^2(\Omega)} \|v\|_{H^1(\Omega)} \leq c_{div}^2 \|\text{dev } \tau\|_{L^2(\Omega)} \|\text{tr}(\tau)\|_{L^2(\Omega)}.$$

Thus we bounded the trace of τ by its deviator, and may conclude

$$a(\tau, \tau) = \frac{1}{2\bar{\mu}} \|\text{dev } \tau\|_{L^2(\Omega)}^2 + \frac{1}{2\bar{\mu}(d\bar{\lambda} + 2\bar{\mu})} \|\text{tr}(\tau)\|_{L^2(\Omega)}^2$$
$$\geq \frac{1}{4\bar{\mu}} \left(\|\text{dev } \tau\|_{L^2(\Omega)}^2 + \frac{1}{c_{div}^2} \|\text{tr}(\tau)\|_{L^2(\Omega)}^2 \right) + 0$$
$$\geq \frac{1}{4\bar{\mu}} \min(1, 1/c_{div}^2) \|\tau\|_\Sigma^2.$$

□

Chapter 4

Finite Element Methods

In this chapter, we concentrate on finite element approximations for the different mixed formulations of elasticity. We shortly recall some basic concepts on Galerkin finite element methods, stability issues and approximation results. Then we review on existing finite element pairs for the discretization of the primal mixed formulation, as well as the Hellinger-Reissner and weak symmetry formulations.

We provide a suitable pair of finite elements for the discretization of our new mixed formulation. There, the displacement space will be approximated using Nédélec finite elements, while we construct a family of tensor-valued, symmetric, normal-normal continuous finite elements for the stresses. We show stability in the discrete setting, using mesh-dependent, broken norms. To be able to give a-priori error estimates, we use interpolation operators. Such operators are well known for the Nédélec space, we construct one for the normal-normal continuous stress space.

Throughout this chapter, we will assume that all essential boundary conditions are homogenous, such that no homogenization of the mixed problem is necessary, and we have $\Sigma_N = \Sigma_0$ and $V_D = V_0$. We further assume that the compliance tensor $\bar{A}$ is well conditioned. In Chapter 5, we will then provide a method which is also stable in case of nearly incompressible materials, when the minimal eigenvalue of $\bar{A}$ tends to zero.

This chapter is organized as follows: In Section 4.1, we are concerned with general techniques and results for mixed finite element methods. Section 4.2 deals with finite element methods for elasticity: First, standard methods are recalled, then a pair of finite elements for our new formulation is introduced. In Section 4.3, this method is analyzed with respect to its stability and approximation properties. Explicit bases for the stress and displacement finite element on a triangular and a tetrahedral reference element are provided in Section 4.4.

4.1 Basic ingredients

We recall basic properties of Galerkin approximations to saddle point problems, concerning stability and a-priori error estimates. We provide the concept of a finite element space on a shape-regular mesh.

We first dwell on the well known abstract framework of Galerkin methods for saddle point problems. We move along the same lines as [BF91]; all proofs can be found in standard finite element literature, wherever mixed problems are concerned, as for example [BF91, GR86, Bra92, BS02].

Then we discuss basic ingredients for the finite element method, such as the underlying discretization of the domain by some shape regular mesh, the finite element itself and the corresponding finite element space, the reference element and the transformation to an element in the mesh.

Let $V_h \subset V_0$, $\Sigma_h \subset \Sigma_0$ be finite-dimensional subspaces of V_0, Σ_0. By using an index h, we indicate that later on the spaces are based on a shape-regular triangulation of the domain Ω of mesh size h. So far, we only assume that the respective families approximate the infinite-dimensional ones, $\bigcup_{h>0} \Sigma_h = \Sigma_0$ and $\bigcup_{h>0} V_h = V_0$. The orthogonal projection operators $\Pi_h^\Sigma : \Sigma_0 \to \Sigma_h$, $\Pi_h^V : V_0 \to V_h$ are naturally defined via

$$\langle \Pi_h^\Sigma \sigma, \tau_h \rangle_\Sigma = \langle \sigma, \tau_h \rangle_\Sigma \quad \forall \tau_h \in \Sigma_h, \tag{4.1}$$
$$\langle \Pi_h^V u, v_h \rangle_V = \langle u, v_h \rangle_V \quad \forall v_h \in V_h. \tag{4.2}$$

We first state the Galerkin approximation of the mixed formulation in Problem 3.21.

Problem 4.1. *Find $\sigma_h \in \Sigma_h$, $u_h \in V_h$ such that*

$$\begin{aligned} a(\sigma_h, \tau_h) + b(\tau_h, u_h) &= \langle F_1, \tau_h \rangle_\Sigma & \forall \tau_h \in \Sigma_h, \\ b(\sigma_h, v_h) &= \langle F_2, v_h \rangle_V & \forall v_h \in V_h, \end{aligned} \tag{4.3}$$

where $a : \Sigma_0 \times \Sigma_0$, $b : \Sigma_0 \times V_0$ are bilinear forms, and F_1, F_2 lie in the respective dual spaces Σ_0^*, V_0^*.

As in the continuous case, we introduce corresponding operators $A_h : \Sigma_h \to \Sigma_h^*$, $B_h : \Sigma_h \to V_h^*$ by

$$\begin{aligned} \langle A_h \sigma_h, \tau_h \rangle_\Sigma &= a(\sigma_h, \tau_h) & \forall \sigma_h, \tau_h \in \Sigma_h, \\ \langle B_h \tau_h, v_h \rangle_V &= b(\tau_h, v_h) & \forall \tau \in \Sigma_h, v \in V_h. \end{aligned}$$

The operator B_h can also be extended to an operator mapping Σ_h to V^* by

$$\langle B_h \tau_h, v \rangle_V := \langle B_h \tau_h, \Pi_h^V v \rangle_V \quad \forall v \in V.$$

4.1.1 Galerkin approximations of saddle point problems

In the following, we recall well-known results on stability and error analysis for the Galerkin approximation in Problem 4.1 of the variational equation in Problem 3.21. We will employ the general framework provided by Brezzi's theory, see [Bre74, BF91]. We assume that the bilinear forms $a(\cdot, \cdot), b(\cdot, \cdot)$ satisfy conditions similar to Assumption 3.22, but on the discrete level.

Assumption 4.2. Let Σ_h, V_h be Hilbert spaces as described above. Let $a : \Sigma_h \times \Sigma_h \to \mathbb{R}$, $b : \Sigma_h \times V_h \to \mathbb{R}$ be bilinear forms with associated discrete operators A_h, B_h satisfying

1. **boundedness of** $a(\cdot,\cdot), b(\cdot,\cdot)$: there exist positive constants $\tilde{c}_{a,2}, \tilde{c}_{b,2} > 0$ such that

$$a(\sigma_h, \tau_h) \leq \tilde{c}_{a,2}\|\sigma_h\|_{\Sigma_h}\|\tau_h\|_{\Sigma_h} \qquad \forall \sigma_h, \tau_h \in \Sigma_h, \tag{4.4}$$
$$b(\tau_h, v_h) \leq \tilde{c}_{b,2}\|\tau_h\|_{\Sigma_h}\|v_h\|_{V_h} \qquad \forall \tau_h \in \Sigma_h, v_h \in V_h, \tag{4.5}$$

2. **discrete inf-sup stability of** $b(\cdot,\cdot)$: there exists a positive constant $\tilde{c}_{b,1} > 0$ independent of h such that

$$\inf_{v_h \in V_h} \sup_{\tau_h \in \Sigma_h} \frac{b(\tau_h, v_h)}{\|\tau_h\|_{\Sigma_h}\|v_h\|_{V_h}} \geq \tilde{c}_{b,1}, \tag{4.6}$$

3. **coercivity of** $a(\cdot,\cdot)$ **on the kernel of** B_h: there exists a positive constant $\tilde{c}_{a,1} > 0$ independent of h such that

$$a(\tau_h, \tau_h) \geq \tilde{c}_{a,1}\|\tau_h\|_{\Sigma_h}^2 \qquad \forall \tau_h \in \operatorname{Ker} B_h, \tag{4.7}$$

where

$$\operatorname{Ker} B_h := \{\tau_h \in \Sigma_h : b(\tau_h, v_h) = 0 \; \forall v_h \in V_h \}.$$

If these assumptions are satisfied for some pair of spaces $\Sigma_h \times V_h$, we can directly conclude existence and uniqueness of a solution to the discrete equations in Problem 4.1, by using Theorem 3.23. For the stability, it is important that the constants $\tilde{c}_{b,1}, \tilde{c}_{a,1}$ do not deteriorate as $h \to 0$.

Note that these conditions are not automatically satisfied for any pair of subspaces $\Sigma_h \times V_h$, even if Assumption 3.22 is met on the infinite-dimensional level. This is due to the fact that, in general, $B_h : \Sigma_h \to V^*$ does not coincide with the restriction of B to Σ_h, as

$$\langle B_h \tau_h, v \rangle_V = \langle B_h \tau_h, \Pi_h^V v \rangle_V = \langle B \tau_h, \Pi_h^V v \rangle_V \neq \langle B \tau_h, v \rangle_V.$$

Thus, also the kernels of B and B_h do not coincide, and both inf-sup stability and coercivity have to be verified in the discrete setting separately. A popular tool to achieve this, is the construction of a *Fortin operator*, which maps the infinite-dimensional space Σ_0 to the subspace Σ_h in a way, that this mapping is orthogonal to V_h with respect to the bilinear form $b(\cdot,\cdot)$.

Let now $(\sigma, u) \in \Sigma_0 \times V_0$ be the solution to the mixed equations (3.18), while $(\sigma_h, u_h) \in \Sigma_h \times V_h$ shall denote the solution to the corresponding discrete system (4.3). We recall well-known a-priori error estimates, as can be found in [BF91, GR86, BS02].

Lemma 4.3. Assume $\Sigma_h \subset \Sigma_0$, $V_h \subset V_0$ are Hilbert spaces. Let $a(\cdot,\cdot)$, $b(\cdot,\cdot)$ satisfy Assumption 4.2, and let (σ_h, u_h) denote the solution to the discrete problem (4.3). Assuming that a solution (σ, u) to the respective continuous problem exists, there hold the error bounds

$$\|\sigma - \sigma_h\|_{\Sigma_h} \leq \left(1 + \frac{\tilde{c}_{a,2}}{\tilde{c}_{a,1}}\right)\left(1 + \frac{\tilde{c}_{b,2}}{\tilde{c}_{b,1}}\right) \inf_{\tau_h \in \Sigma_h} \|\sigma - \tau_h\|_{\Sigma_h} + \frac{\tilde{c}_{b,2}}{\tilde{c}_{a,1}} \inf_{v_h \in V_h} \|u - v_h\|_{V_h},$$

$$\|u - u_h\|_{V_h} \leq \left(1 + \frac{\tilde{c}_{b,2}}{\tilde{c}_{b,1}}\right) \inf_{v_h \in V_h} \|u - v_h\|_{V_h} + \frac{\tilde{c}_{a,2}}{\tilde{c}_{b,1}} \|\sigma - \sigma_h\|_{\Sigma_h}.$$

If the finite element spaces Σ_h, V_h are chosen such that $\operatorname{Ker} B_h \subseteq \operatorname{Ker} B$, there holds the improved error bound

$$\|\sigma - \sigma_h\|_{\Sigma_h} \leq \left(1 + \frac{\tilde{c}_{a,2}}{\tilde{c}_{a,1}}\right)\left(1 + \frac{\tilde{c}_{b,2}}{\tilde{c}_{b,1}}\right) \inf_{\tau_h \in \Sigma_h} \|\sigma - \tau_h\|_{\Sigma_h}.$$

4.1.2 Triangulation

So far, we only assumed that Σ_h, V_h are families of finite dimensional spaces approximating the infinite dimensional spaces Σ_0, V_0. In many applications, these finite element spaces are derived from a family of underlying discretizations of the domain Ω. The parameter h is then associated to the mesh size of these triangulations, the method is referred to as "h-FEM". We now provide basic concepts for such families of triangulations. We assume the domain $\Omega \subset \mathbb{R}^d, d = 2, 3$ to have a piecewise polynomial boundary which is Lipschitz continuous.

Definition 4.4 (Regular Triangulation). *Let $\mathcal{T} := \{T\}$ be a non-overlapping domain decomposition of Ω into elements T of simple geometry. We call $\mathcal{T}$ a regular triangulation of Ω iff*

1. *the elements are non-overlapping, $T \cap \tilde{T} = \emptyset$ for $T \neq \tilde{T} \in \mathcal{T}$,*
2. *the domain Ω is covered by the triangulation, $\bar{\Omega} = \bigcup_{T \in \mathcal{T}} \bar{T}$,*
3. *the intersection of two elements is either empty, or a common edge, face, or vertex of both.*

We further call the triangulation simplicial, *if all elements are simplices, i.e triangles for $d = 2$ and tetrahedrons for $d = 3$.*

Note that condition 3. excludes hanging nodes. We introduce the set of vertices $\mathcal{V} = \{V\}$, the set of element edges $\mathcal{E} = \{E\}$, and the set of element interfaces or facets $\mathcal{F} = \{F\}$. For two-dimensional meshes, the last two sets coincide. Sometimes, it will not be necessary to distinguish between edges, facets and elements. Then we use the union

$$\mathcal{X} = \mathcal{E} \cup \mathcal{F} \cup \mathcal{T},$$

and mean by $X \in \mathcal{X}$ any edge, facet or element in the triangulation. We will further use $\mathcal{E}(T) := \{E \in \mathcal{E} : E \subset \partial T\}$ and similarly $\mathcal{F}(T)$ as the set of edges/facets corresponding to element T. The set $\mathcal{X}(T)$ is again the union of edges, facets, and the element itself; $\mathcal{X}(T) = \mathcal{E}(T) \cup \mathcal{F}(T) \cup \{T\}$. Furthermore, for some element T, we define Δ_T to be the patch of all neighboring elements, i.e.

elements sharing at least a vertex with T. Similarly, Δ_E, Δ_F shall denote the patches of neighboring elements for an edge E or a facet F.

We assume that all facets $F \in \mathcal{F}$ are oriented, i.e. that their normal n_F is uniquely determined. This allows the following definition of jump operators.

Definition 4.5 (jump operators). *For a facet $F \in \mathcal{F}$, let T_1, T_2 be its neighboring elements, where we assume that $n_F = n_{T_1} = -n_{T_2}$. Let $w : (T_1 \cup T_2) \to \mathbb{R}$ be piecewise smooth, then the jump operator is defined by*

$$[\![w]\!]_F := w|_{T_1} - w|_{T_2}. \tag{4.8}$$

For a vector-valued function $v : (T_1 \cup T_2) \to \mathbb{R}^d$, the normal jump is given by

$$[\![v]\!]_{n,F} := v_{n_{T_1}}|_{T_1} + v_{n_{T_2}}|_{T_2} = [\![v_n]\!]_F. \tag{4.9}$$

For a boundary facet $F \subset (\partial T \cap \Gamma)$, the jump and normal jump are defined by

$$[\![w]\!]_F := w|_T, \qquad [\![v]\!]_{n,F} := v_n \tag{4.10}$$

The jump operator can be used to impose continuity of a piecewise defined function, as well as homogenous boundary conditions. If the facet F is clear from context, the subscript may also be omitted, we use then $[\![\cdot]\!]$, $[\![\cdot]\!]_n$ only.

In the finite element method, it is common to define basis functions of the respective discrete spaces not directly for the global mesh, but on the *reference element*. They are then mapped to an element in the mesh using a suitable transformation. The reference element is usually of unit size, and of the same topological shape as the element. We will consider unit segments in 1D, unit triangles or squares in 2D, and tetrahedrons and prisms in 3D as reference elements. Also numerical integration or calculation of derivatives can be done on the reference element, which is useful especially in case of curved elements.

Let now $\hat{T}$ be the reference element for some element $T \in \mathcal{T}$. Let

$$\Phi_T : \hat{T} \to T, \hat{x} \mapsto \Phi_T(\hat{x}) =: x$$

a smooth, one-to-one mapping from the reference element $\hat{T}$ to an element T. We use the convention to mark all quantities on the reference element by a hat. A point $\hat{x} \in \hat{T}$ is then mapped to $x \in T$.

The Jacobian of this transformation shall be denoted by F_T,

$$F_T(\hat{x}) = \left(\frac{\partial \Phi_{T,i}}{\partial \hat{x}_j}(\hat{x}) \right)_{ij}.$$

The Hessian of the i-th component H_T^i is given by

$$H_T^i(\hat{x}) = \left(\frac{\partial^2 \Phi_{T,i}}{\partial \hat{x}_j \partial \hat{x}_k}(\hat{x}) \right)_{jk}, \quad 1 \leq i, j, k \leq d.$$

The Jacobi determinant is given by $J_T(\hat{x}) = \det(F_T(\hat{x}))$. Similarly, for a facet F or an edge E, let J_F, J_E be the transformation of measures of the facet transformation $\hat{F} \to F$ or the edge transformation $\hat{E} \to E$ respectively. Now, let E, F be an edge, facet of element $T \in \mathcal{T}$, respectively. For a normal vector n on F and a tangential vector τ along E we have

$$n = J_T/J_F \, F_T^{-T} \hat{n}, \quad \tau = 1/J_E \, F_T \hat{\tau}.$$

In general, the map Φ_T is non-linear. We will use polynomial mappings, such that polynomial boundaries can be represented exactly. In case of an affine linear map Φ_T, the Jacobian F_T is constant on T, and the Hessians $H_T^i, i = 1, \ldots d$ vanish. Such a transformation is used for simplicial elements or elements of tensor product structure, as long as no curvature of the elements is necessary. In our analysis, we will only consider affine linear transformations, if not indicated differently. Then, the Jacobi determinants J_T, J_F, J_E of element, facet, or edge transformation equal the respective relative sizes of T, F and E,

$$J_T = \frac{|T|}{|\hat{T}|}, \quad J_F = \frac{|F|}{|\hat{F}|}, \quad J_E = \frac{|E|}{|\hat{E}|}.$$

Using Φ_T, we can define the local mesh size for some $x \in T, x = \Phi_T(\hat{x})$

$$h(x) := |F_T(\hat{x})|_s, \quad h_T := \max_{x \in T} h(x).$$

Here we use the spectral norm $|\cdot|_s$ of the matrix-valued quantity F_T. For a facet F, let T_1, T_2 be the adjacent elements. Then we set $h_F := (|T_1| + |T_2|)/|F|$ as the average height of those two elements perpendicular to F.

Last, we define the notion of quasi-uniform and shape-regular triangulations.

Definition 4.6. Let $(\mathcal{T}_h)_{h \to 0}$ be a family of regular triangulations of Ω, where $\mathcal{T}_h$ is of maximum mesh size $h = \max_{T \in \mathcal{T}_h} h_T$. Let $\mathcal{F}_h, \mathcal{E}_h,$ and $\mathcal{V}_h$ denote the respective sets of facets, edges, and vertices of the triangulation $\mathcal{T}_h$.

1. The family $(\mathcal{T}_h)$ is called **uniform** iff there exists a global mesh size h such that $h(x) = h$ for all $x \in \Omega$.

2. The family $(\mathcal{T}_h)$ is called **quasi-uniform** iff for each element $T \in \mathcal{T}_h$ the local mesh size h_T is proportional to a global mesh size h,

$$c_1 h \leq h_T \leq c_2 h \quad \forall T \in \mathcal{T}_h,$$

where c_1, c_2 are independent of h.

3. The family $(\mathcal{T}_h)$ is called **shape-regular** iff there exist constants $c_1, c_2 > 0$ independent of the

local mesh size h_T such that

$$c_1 \leq \|F_T\|\|F_T^{-1}\| \leq c_2 \quad \forall T \in \mathcal{T}_h, \ h > 0.$$

4.1.3 Finite element spaces

We now define the notion of finite element spaces, which are based on a triangulation of the computational domain. Therefore, each element in the triangulation is equipped with a *finite element*. The union of these finite elements is then used to build the global finite element space. Following [BS02], we use the following definition of a finite element due to [Cia78].

Definition 4.7 (Finite Element). *A finite element is a triplet* $(T, X, \mathcal{N})$ *consisting of*

1. *the element domain* $T \subset \mathbb{R}^d$, *which is a bounded closed set with non-empty interior and piecewise smooth boundary,*

2. *the space of shape functions* X, *which is of finite dimension,*

3. *the set of nodal variables or degrees of freedom (dofs)* $\mathcal{N} = \{N_1, \ldots N_k\}$, *which is a basis for* X^*.

In order to specify the space of shape functions X, one can use any set of linear independent functions spanning X. For a given set of nodal variables $\mathcal{N}$, there exists a unique nodal basis.

Definition 4.8 (Nodal basis). *Let* $(T, X, \mathcal{N})$ *be a finite element. The basis* $\{\varphi_i, i = 1, \ldots, k\}$ *is called a* nodal basis *for* X *if it is dual to* $\mathcal{N}$, *i.e.*

$$N_i(\varphi_j) = \delta_{ij} \quad \text{for } i, j = 1, \ldots, k.$$

Inspired by [Cia91], we define a finite element space.

Definition 4.9. *Let now* $\mathcal{T}_h$ *be a regular triangulation of the domain* $\Omega \subset \mathbb{R}^d$, *where each* $T \in \mathcal{T}_h$ *is equipped with a finite element* $(T, X(T), \mathcal{N}(T))$. *The associated finite element space* X_h *is defined by*

$$X_h := \left\{ v \in \Pi_{T \in \mathcal{T}_h} X(T) : N(v|_{T_i}) = N(v|_{T_j}) \text{ for all } N \in \mathcal{N}(T_i) \cap \mathcal{N}(T_j) \right\}$$

as the space of finite element shape functions, where degrees of freedom shared between elements coincide.

Using the nodal basis, it is then straightforward to define the local and global *nodal interpolation operators*.

Definition 4.10 (Nodal interpolation operator).

1. Consider a finite element $(T, X(T), \mathcal{N}(T))$ with nodal basis $\{\varphi_i : T \to \mathcal{R}, i = 1, \ldots k\}$ for some finite dimensional space $\mathcal{R}$. Let $v : T \to \mathcal{R}$ be a sufficiently smooth map such that $N_i(v)$ is well defined for $i = 1, \ldots k$. Then we call

$$\mathcal{I}^{X(T)} v := \sum_{i=1}^{k} N_i(v) \varphi_i$$

   the *local nodal interpolant of v*.

2. Let now $\mathcal{T}_h$ be a regular triangulation of the domain $\Omega \subset \mathbb{R}^d$, where each $T \in \mathcal{T}_h$ is associated with a finite element $(T, X(T), \mathcal{N}(T))$. For $v : \Omega \to \mathcal{C}$ sufficiently smooth, we define the global nodal interpolation operator $\mathcal{I}_h^X$ mapping to the associated finite element space X_h by

$$(\mathcal{I}_h^X v)|_T := \mathcal{I}^{X(T)}(v|_T) \qquad \forall T \in \mathcal{T}_h.$$

4.2 Finite element methods for mixed elasticity

We shortly comment on existing families of finite elements for mixed elasticity. Then, we introduce finite elements for our approach. We use Nédélec finite elements (edge elements) for the displacements, as they are a standard choice when discretizing $H(\mathrm{curl})$. We refer to [Néd80, Néd86] for their first introduction. There the tangential components along edges and across faces are continuous. For the stresses, we introduce a new finite element space. Its degrees of freedom enforce normal-normal continuity across element interfaces. We prove existence and uniqueness of the discrete solution, as well as its continuous dependence on the given data. Using nodal interpolation operators, we provide a-priori error estimates.

Throughout the following, we assume that all boundary conditions are homogenous, i.e.

$$u_D = 0 \text{ on } \Gamma_D, \qquad \vec{t}_N = 0 \text{ on } \Gamma_N.$$

Thus, the spaces Σ_0, Σ_N and V_0, V_D coincide. Inhomogeneous boundary conditions can be included into the variational formulations and finite element spaces in a straightforward way.

Moreover, we assume that Ω is a polyhedral Lipschitz domain, which is decomposed by a family of shape-regular, quasi-uniform affine linear triangulations $(\mathcal{T}_h)$ of maximum mesh size h. For our constructions, we will further assume that the triangulation $\mathcal{T}_h$ is simplicial, i.e. it consists of triangular elements in 2D, or tetrahedral elements in 3D. In Chapter 6, we extend these results to tensor product elements, such as quadrilaterals or prisms.

Notation All our finite element spaces will be based on piecewise polynomial functions on a triangulation $\mathcal{T}_h$. On some simplicial element T, we use the space $P^k(T)$, which consists of polynomials up to order k. Similarly, $P^k_{SYM}(T)$ shall be the space of tensor-valued symmetric functions where each component lies in $P^k(T)$. We write $P^k(\mathcal{T}), P^k_{SYM}(\mathcal{T})$ for the respective piecewise defined spaces, without any continuity assumptions across element interfaces. We additionally need the "bubble space" $P^k_0(T)$, which is the subset of $P^k(T)$ of functions satisfying homogenous Dirichlet boundary conditions on ∂T. Furthermore, we denote $P^k_{0,\tau}(T)$ to be the space of vector-valued polynomial functions with vanishing tangential trace. Similarly, $P^k_{0,nn}(T)$ shall be the space of tensor-valued symmetric polynomials, where the normal-normal component is zero on the boundary. In short, we use the following polynomial spaces satisfying boundary conditions

$$\begin{aligned} P^k_0(T) &:= \{q \in P^k(T) : q|_{\partial T} = 0\}, \\ P^k_{0,\tau}(T) &:= \{q \in [P^k(T)]^d : q_\tau|_{\partial T} = 0\}, \\ P^k_{0,nn}(T) &:= \{q \in P^k_{SYM}(T) : q_{nn}|_{\partial T} = 0\}. \end{aligned}$$

We will dwell on the explicit construction of bases for all these spaces in Section 4.4.

4.2.1 Existing methods

We shortly outline possible methods of discretization for the primal, the Hellinger-Reissner, and the mixed method with weak symmetry.

4.2.1.1 FEM for the primal method

When discretizing the primal mixed method with finite elements of order k, the most natural choice is to use the spaces

$$\Sigma^k_{pr,h} := \{\tau_h \in L^2_{SYM}(\Omega) : \tau_h \in P^k_{SYM}(\mathcal{T}_h)\}, \tag{4.11}$$

$$V^k_{pr,h} := \{v_h \in [L^2(\Omega)]^d : v_h \in P^{k+1}(\mathcal{T}_h),\ v_h \text{ cont.},\ v_h = 0 \text{ on } \Gamma_D\}. \tag{4.12}$$

These spaces satisfy an inf-sup condition, as for each $v_h \in V^k_{pr,h}$ the tensor $\tau_h := \varepsilon(v_h)$ lies in $\Sigma^k_{pr,h}$, and therefore

$$\inf_{v_h \in V^k_{h,pr}} \sup_{\tau_h \in \Sigma^k_{pr,h}} b(v_h, \tau_h) \geq \int_\Omega |\varepsilon(v_h)|^2\,dx \geq \tilde{c}_K c_F \|v_h\|_{H^1(\Omega)} \|\tau_h\|_{L^2(\Omega)}.$$

The constant of stability again depends on the constant in Korn's inequality, therefore we expect locking when treating anisotropic geometries and/or meshes. Also, the constant of coercivity for $a(\cdot, \cdot)$ depends on the smallest eigenvalue of $\bar{A}$, which tends to zero for almost incompressible materials. Thus, the method (and especially the equivalent method based on the pure displacement formulation), is a good choice as long as the above problems do not occur. Finite element methods based on the pure displacement formulation are probably the fastest methods available.

4.2.1.2 FEM for the Hellinger-Reissner formulation

A main achievement of the Hellinger-Reissner formulation is the fact that, in the continuous setting, it is stable independently of the Lamé parameter $\bar{\lambda}$ going to infinity. To discretize the Hellinger-Reissner formulation, one needs to construct a pair of finite element spaces $\Sigma_{HR,h}^k \times V_{HR,h}^k$, such that they are conforming for the infinite dimensional spaces $H_{SYM}(\text{div})$ and $[L^2(\Omega)]^d$. The latter space can simply be set up using piecewise polynomial functions of some prescribed degree. However, finite elements for the stress space need

- to be symmetric tensor-valued,
- to be normal continuous
- to satisfy $\text{Ker } B_h \subset \text{Ker } B$, or $\text{div } \Sigma_{HR,h}^k \subset V_{HR,h}^k$.

A construction of such spaces cannot be done in a simple way. In three space dimensions, a rigorous analysis of these spaces was done in [AC05], a finite element basis in two and three space dimensions is provided in [AW02, AAW08]. In the plane, the lowest order stress element consists of piecewise cubics with linear divergence, whereas the displacement is approximated by discontinuous, piecewise linear finite elements. In three dimensions, the stress element has to contain the full quadratic space, enhanced by divergence free functions of orders 3 and 4. Again, the displacement space consists of piecewise linears. The local dimension of the stress space is then 24 in two, and 162 in three dimensions.

4.2.1.3 FEM for methods with weakly imposed symmetry

When discretizing methods with weak symmetry, one does not need to use stress finite elements which are symmetric. Therefore, one can use well-known elements for $H(\text{div})$, and take them d times to get an approximation of $[H(\text{div})]^d$. Many different choices of triples of finite elements have been analyzed by several authors, not all of these triples are conforming. We refer to [Ste86, Ste88, AFW07]. Probably the best-known among these elements is the PEERS element [ABD84], which was also analyzed in the general framework of [AFW07]. There, one chooses the famous Raviart-Thomas finite element space $[RT_0]^d$ for the stresses, augmented by curls of bubble functions. For a definition of the Raviart-Thomas space, see [RT77, Néd80]. It consists of vector-valued linear functions, which have constant normal components on the facets of a triangle/tetrahedron. Then, the displacement is approximated using piecewise constants, and piecewise linear, continuous functions are used for the Lagrangian multipliers, which may be interpreted as the rotation of the displacement. These elements are stable also for nearly incompressible materials. Counting the coupling degrees of freedom of a tetrahedral element, one obtains 24 nodal values for a method of approximation order one. Increasing all orders by one, we get 66 coupling nodal values per element for a second order method.

4.2.2 The TD-NNS method

We are concerned with discretizing the infinite dimensional set of equations (3.35), (3.36) by suitable finite element spaces Σ_h approximating Σ_0, V_h approximating V_0. As our choice for the displacement space was $V = H(\text{curl})$, we use tangential continuous finite elements in order to be conforming. For $\Sigma = H(\text{div div})$, we saw the need of continuous normal-normal components of the stress tensor. Note that this condition alone does not imply conformity for a piecewise smooth function τ, not even for a piecewise polynomial: Additionally, the local traces of the normal-tangential component $\tau_{n\tau}$ has to lie in $H_\|^{-1/2}(\text{div}_{\partial T})$, cf. Remark 3.37. This is a local continuity constraint on $\tau_{n\tau}$ which is not satisfied in general. However, we violate this condition in the proposed finite element spaces, which is rectified by the analysis provided.

In several occasions, we also need a finite element space W_h^k which approximates the Sobolev space $H_{0,\Gamma_D}^1(\Omega)$ conformingly, and the space $\mathcal{P}_k^h$ consisting of piecewise polynomial functions without any continuity assumptions. We use the spaces

$$\Sigma_h^k := \{\tau_h \in L_{SYM}^2(\Omega) : \tau_h \in P_{SYM}^k(\mathcal{T}_h),\ \tau_{h,nn}\ \text{cont.},\ \tau_{h,nn} = 0 \text{ on } \Gamma_N\}, \quad (4.13)$$

$$V_h^k := \{v_h \in [L^2(\Omega)]^d : v_h \in [P^k(\mathcal{T}_h)]^d,\ v_{h,\tau}\ \text{cont.},\ v_{h,\tau} = 0 \text{ on } \Gamma_D\}, \quad (4.14)$$

$$W_h^k := \{w_h \in \mathcal{C}(\Omega) : w_h \in P^k(\mathcal{T}_h),\ w_h = 0 \text{ on } \Gamma_D\}, \quad (4.15)$$

$$\mathcal{P}_h^k := \{q_h \in L^2(\Omega) : q_h \in P^k(\mathcal{T}_h)\}. \quad (4.16)$$

There, and throughout the remainder of this work, $k \in \mathbb{N}_0$ shall be a non-negative integer. We may omit it as an index, if it is not necessary in the context. The above choice of spaces induces the following mixed problem formulation.

Problem 4.11 (TD-NNS formulation). *Find $(\sigma_h, u_h) \in \Sigma_h^k \times V_h^k$ defined as above such that*

$$\begin{aligned} a(\sigma_h, \tau_h) + b(\tau_h, u_h) &= \langle F_1, \tau_h \rangle_\Sigma & \forall \tau_h \in \Sigma_h^k, \\ b(\sigma_h, v_h) &= \langle F_2, v_h \rangle_V & \forall v_h \in V_h^k. \end{aligned} \quad (4.17)$$

The bilinear forms $a(\cdot,\cdot), b(\cdot,\cdot)$ and the right hand sides F_1, F_2 are defined by the relations (3.37) – (3.40). Due to the smoothness of finite element functions, one can evaluate $b(\cdot,\cdot)$ using

$$b(\tau_h, v_h) = \langle \text{div } \tau_h, v_h \rangle_V = \sum_{T \in \mathcal{T}_h} \left[\int_T \text{div } \tau_h \cdot v_h \, dx - \int_{\partial T} \tau_{h,n\tau} \cdot v_{h,\tau} \, ds \right] \quad (4.18)$$

$$= \sum_{T \in \mathcal{T}_h} \left[-\int_T \tau_h : \varepsilon(v_h) \, dx + \int_{\partial T} \tau_{h,nn} v_{h,n} \, ds \right]. \quad (4.19)$$

The evaluation of $b(\cdot,\cdot)$ by (4.18), (4.19) for piecewise smooth functions τ_h, v_h can be seen easily: Let first v be a (globally) smooth test function. with tangential component vanishing on

Γ_D. The divergence operator div for general $\tau \in H(\text{div div})$ is defined as

$$\langle \text{div } \tau, v \rangle_V := -\int_\Omega \tau : \varepsilon(v) \, dx + \int_{\Gamma_D} \tau_{nn} v_n \, ds,$$

where the surface integral is understood as duality product. For the piecewise smooth function τ_h, we may integrate by parts on each element,

$$\begin{aligned}\langle \text{div } \tau_h, v \rangle_V &= \sum_{T \in \mathcal{T}_h} \left[\int_T \text{div } \tau_h \cdot v \, dx - \int_{\partial T} \tau_{h,n} \cdot v \, ds \right] + \int_{\Gamma_D} \tau_{h,nn} v_n \, ds \\ &= \sum_{T \in \mathcal{T}_h} \left[\int_T \text{div } \tau_h \cdot v \, dx - \int_{\partial T} (\tau_{h,n\tau} \cdot v_\tau + \tau_{h,nn} v_n) \, ds \right] + \int_{\Gamma_D} \tau_{h,nn} v_n \, ds.\end{aligned}$$

Reordering the boundary terms for the normal components facet by facet, we see

$$\begin{aligned}\langle \text{div } \tau_h, v \rangle_V &= \sum_{T \in \mathcal{T}_h} \left[\int_T \text{div } \tau_h \cdot v \, dx - \int_{\partial T} \tau_{h,n\tau} \cdot v_\tau \, ds \right] - \sum_{\substack{F \in \mathcal{F}_h \\ F \subset \Omega}} \int_F \underbrace{[\![\tau_{h,nn} v_n]\!]}_{=0} \, ds \\ &+ \int_{\Gamma_D} \underbrace{(\tau_{h,nn} v_n - \tau_{h,nn} v_n)}_{=0} \, ds - \int_{\Gamma_N} \underbrace{\tau_{h,nn}}_{=0} v_n \, ds.\end{aligned}$$

The jump terms for interior facets cancel out, as $\tau_h \in \Sigma_h^k$ is normal-normal continuous, and v is smooth. Similarly, on the Dirichlet boundary Γ_D, the terms cancel, whereas $\tau_{h,nn}$ is zero on the remaining part Γ_N. This yields equality (4.18), the second identity (4.19) follows by integration by parts.

We give a more detailed description of the spaces Σ_h^k, V_h^k, and W_h^k and the underlying finite elements in the sequel. We will see that the lowest order method is given for $k = 1$, and results in 24 coupling degrees of freedom on a tetrahedron: $3 \cdot 4 = 12$ for the piecewise linear normal-normal stresses on the facets, plus $2 \cdot 6 = 12$ for the piecewise linear tangential displacements on element edges. This is the same number as for the PEERS element; however, we propose a second order method which needs again 12 coupling dofs for the stresses, and 30 for the displacements: this results in a total number of 42, which is considerably lower than the 66 dofs, which one obtains for the PEERS element.

4.2.2.1 Finite elements for H^1

High order finite elements for H^1 have been constructed in many different ways: The most straightforward idea is to use point evaluations as nodal values, where the points are distributed regularly, see e.g. [QV97, Cia91]. These elements are not often used in high order computations, as their numerical stability for higher polynomial orders deteriorate. Spectral elements [KS99] are more suitable for high order finite elements. In this work, we use a hierarchical basis for the discretization of H^1. A first general construction of such elements was given in [AC03]. We stay close to the notation and results from [Zag06]. There, the order of the finite element can be chosen for each edge, face, and for the interior separately. However, we will assume uniform polynomial order on the whole space, as this very much simplifies notation.

For $T \in \mathcal{T}_h$, we define the H^1 conforming element of order $k \in \mathbb{N}$ as the triplet $(T, W^k(T), \mathcal{N}_k^W(T))$. We use the full polynomial space $W^k(T) := P^k(T)$. The set of nodal values $\mathcal{N}_k^W(T)$ consists of vertex-, edge-, facet- and cell-based degrees of freedom,

$$\mathcal{N}_k^W(T) := \left(\bigcup_{V \in \mathcal{V}(T)} \mathcal{N}_V^W \right) \cup \left(\bigcup_{E \in \mathcal{E}(T)} \mathcal{N}_{E,k}^W \right) \cup \left(\bigcup_{F \in \mathcal{F}(T)} \mathcal{N}_{F,k}^W \right) \cup \mathcal{N}_{T,k}^W.$$

The respective sets are defined via

- for $V \in \mathcal{V}(T)$, $\mathcal{N}_V^W := \{N_V^W\}$ where

$$N_V^W(w) := w(V).$$

- for $E \in \mathcal{E}(T)$, $\mathcal{N}_{E,k}^W := \{N_{E,i}^W : 2 \leq i \leq k\}$ where

$$N_{E,i}^W(w) := \int_E \frac{\partial w}{\partial s} \frac{\partial q_i}{\partial s} \, ds$$

with $\{q_i : 2 \leq i \leq k\}$ a basis for $P_0^k(E)$.

- in 3D, for $F \in \mathcal{F}(T)$, $\mathcal{N}_{F,k}^W := \{N_{F,i}^W\}$ where

$$N_{F,i}^W(w) := \int_F \nabla_F w \cdot \nabla_F q_i \, ds,$$

where $\{q_i\}$ form a basis for $P_0^k(F)$.

- $\mathcal{N}_{T,k}^W := \{N_{T,i}^W\}$ where

$$N_{T,i}^W(v) := \int_T \nabla w \cdot \nabla q_i \, dx,$$

where $\{q_i\}$ form a basis for $P_0^k(T)$.

These degrees of freedom are unisolvent for the local space $W^k(T)$.

4.2.2.2 Finite elements for $H(\text{curl})$

According to Theorem 3.12, for the discretization of $H(\text{curl})$ one needs to use finite elements, whose degrees of freedom ensure tangential continuity of the discrete functions. Elements of this kind were first proposed in [Néd80, Néd86]. We use Nédélec elements of variable order of the second kind, as proposed in [SZ05, Zag06]. In this approach, again the order of the finite element can be chosen independently for each edge, facet and the interior. For a compact presentation, we restrict ourselves to a uniform order for the whole space.

We define the Nédélec element of order k associated to an element $T \in \mathcal{T}_h$ as the triple $(T, V^k(T), \mathcal{N}_k^V(T))$. There, $V^k(T) = [P^k(T)]^d$ is the full polynomial space of order at most k. The set of degrees of freedom $\mathcal{N}_k^V(T)$ can be divided into edge-, face- and cell-based dofs,

$$\mathcal{N}_k^V(T) := \left(\bigcup_{E \in \mathcal{E}(T)} \mathcal{N}_{E,k}^V \right) \cup \left(\bigcup_{F \in \mathcal{F}(T)} \mathcal{N}_{F,k}^V \right) \cup \mathcal{N}_{T,k}^V.$$

The respective sets are defined via

- for $E \in \mathcal{E}(T)$, $\mathcal{N}_{E,k}^V := \{N_{E,i}^V : 0 \leq i \leq k\}$ where

$$N_{E,i}^V(v) := \int_E v_{\tau_E} q_i \, ds$$

with $\{q_i : 0 \leq i \leq k\}$ a basis for $P^k(E)$.

- in 3D, for $F \in \mathcal{F}(T)$, $\mathcal{N}_{F,k}^V := \{N_{F,i}^V, N_{F,j}^V\}$ where

$$N_{F,i}^V(v) := \int_F \operatorname{curl}_F(v) \cdot \operatorname{curl}_F(q_i) \, ds, \qquad N_{F,j}^V(v) := \int_F v \cdot r_j \, ds$$

where $\{q_i\}$ are such that $\{\operatorname{curl}_F q_i\}$ form a basis for $\operatorname{curl}_F(P_{0,\tau}^k(F))$, and $\{r_j\}$ are a basis for $\nabla_F(P_0^{k+1}(F))$.

- $\mathcal{N}_{T,k}^V := \{N_{T,i}^V, N_{T,j}^V\}$ where

$$N_{T,i}^V(v) := \int_T \operatorname{curl}(v) \cdot \operatorname{curl}(q_i) \, dx, \qquad N_{T,j}^V(v) := \int_T v \cdot r_j \, dx$$

where $\{q_i\}$ are such that $\{\operatorname{curl} q_i\}$ form a basis for $\operatorname{curl}(P_{0,\tau}^k(T))$, and $\{r_j\}$ are a basis for $\nabla(P_0^{k+1}(T))$.

Recall that $P_{0,\tau}^k(T)$ is the space of vector-valued polynomials up to order k, which satisfy homogenous tangential boundary conditions on the boundary ∂T of the simplex T. The degrees of freedom are unisolvent for the local space $V^k(T)$, see [Mon03], where they are defined also for elements of variable order.

4.2.2.3 Finite elements for $H(\operatorname{div} \operatorname{div})$

We provide a suitable finite element for the space Σ_h^k. The degrees of freedom are chosen such that the normal-normal component of a finite element function is continuous across interfaces. We add degrees of freedom interior to the element, such that these nodal variables are unisolvent for the local space.

Our finite element associated to $T \in \mathcal{T}_h$ is given by $(T, \Sigma^k(T), \mathcal{N}_k^\Sigma(T))$. There we define the local finite element space $\Sigma^k(T) := P_{SYM}^k(T)$ as the full polynomial space of symmetric, tensor-valued fields. The set of degrees of freedoms $\mathcal{N}_k^\Sigma(T)$ consists of values associated to facets and cells,

$$\mathcal{N}_k^\Sigma(T) = \left(\bigcup_{F \in \mathcal{F}(T)} \mathcal{N}_{F,k}^\Sigma\right) \cup \mathcal{N}_{T,k}^\Sigma.$$

We define the respective sets

- for $F \in \mathcal{F}(T)$, $\mathcal{N}_{F,k}^\Sigma = \{N_{F,i}^\Sigma : i \in I_F\}$ where

$$N_{F,i}^\Sigma(\tau) := \int_F J_F \, \tau_{nn} q_i \, ds, \qquad (4.20)$$

where $\{q_i, i \in I_F\}$ is a basis for $P^k(F)$, and I_F is a suitable index set,

- $\mathcal{N}_{T,k}^\Sigma = \{N_{T,i}^\Sigma : i \in I_T\}$ where

$$N_{T,i}^\Sigma(\tau) := \int_T J_T \, \tau : (F_T^{-T} \gamma_i F_T^{-1}) \, dx. \qquad (4.21)$$

There $\{\gamma_i : i \in I_T\}$ is a basis for $P_{0,nn}^k(T)$, the space of polynomial symmetric tensor fields up to order k, with homogenous normal-normal boundary conditions, and I_T is a suitable index set.

The following lemma states that these degrees of freedom are unisolvent for the local space $\Sigma^k(T)$; we will give an explicit basis for this space in Section 4.4.

Lemma 4.12. *For $T \in \mathcal{T}_h$, the triple $(T, \Sigma^k(T), \mathcal{N}_k(T))$ defined as above forms a finite element, the nodal variables $\mathcal{N}_k(T)$ are a basis for $\Sigma^k(T)^* = P_{SYM}^k(T)^*$.*

Proof. For $\tau_h \in \Sigma^k(T)$, we show

$$N^\Sigma(\tau_h) = 0 \quad \forall N^\Sigma \in \mathcal{N}_k^\Sigma(T) \qquad \Longrightarrow \qquad \tau_h = 0.$$

From the fact that $N_{F,i}^\Sigma \tau_h = 0$ for all $i \in I_F$ on all boundary facets $F \in \mathcal{F}(T)$, we deduce that $\tau_{h,nn} = 0$ on ∂T. Thus, τ_h lies in $P_{0,nn}^k(T)$, it has a representation in the basis $\{\gamma_i, i \in I_T\}$, $\tau_h = \sum_{i \in I_T} a_i \gamma_i$. We obtain for the degrees of freedom associated to the interior, for all $i \in I_T$

$$\begin{aligned}
0 &= N_{F,i}^\Sigma(\tau_h) = \int_T J_T \Big(\sum_{j \in I_T} a_j \gamma_j\Big) : (F_T^{-T} \gamma_i F_T^{-1}) \, dx \\
&= \sum_{j \in I_T} \int_T J_T a_j (\gamma_j F_T^{-1})(\gamma_i F_T^{-1}) \, dx.
\end{aligned}$$

From this we conclude that $a_i = 0$ for $i \in I_T$, as the $\{\gamma_i : i \in I_T\}$ are linearly independent, F_T is invertible, and $J_T \neq 0$. □

4.2.2.4 Finite elements for L^2

The last finite element we discuss is suitable for the discretization of L^2. The corresponding finite element space is the space $\mathcal{P}_h^k$ consisting of piecewise polynomial functions, without any restrictions on the inter-element continuity. The degrees of freedom correspond to a family of, preferably orthogonal, polynomials (q_i). Our finite element associated to $T \in \mathcal{T}_h$ is given by $(T, \mathcal{P}^k(T), \mathcal{N}_k^{\mathcal{P}}(T))$. The local finite element space $\mathcal{P}^k(T) := P^k(T)$ is the full polynomial space of order up to k. The set of degrees of freedoms $\mathcal{N}_k^{\mathcal{P}}(T)$ consists of cell-bound values only,

$$\mathcal{N}_k^{\mathcal{P}}(T) = \mathcal{N}_{T,k}^{\mathcal{P}}.$$

There, we define

- $\mathcal{N}_{T,k}^{\mathcal{P}} = \{N_{T,i}^{\mathcal{P}}\}$ where

$$N_{T,i}^{\mathcal{P}}(p) := \int_T p\, q_i\, dx, \tag{4.22}$$

where $\{q_i\}$ is a basis for $P^k(T)$.

These degrees of freedom are obviously unisolvent for the local space $\mathcal{P}^k(T)$. We note, that if the family (q_i) is chosen orthogonal in the L^2 sense, it coincides with the nodal basis. The corresponding nodal interpolation operator $\mathcal{I}_k^{\mathcal{P}}$ is the L^2 projection.

4.2.2.5 Transformations

In order to transform functions from the reference element to an element T in the mesh, we need conforming transformations, which preserve the degrees of freedom of the finite element. For an element $T \in \mathcal{T}_h$, let $x = \Phi_T(\hat{x})$ be the point corresponding to $\hat{x} \in \hat{T}$. Let $\hat{\Sigma}^k := \Sigma^k(\hat{T}), \hat{V}^k := V^k(\hat{T})$ and $\hat{W}^k := W^k(\hat{T})$ be the local spaces on the reference element. We introduce the local operators $\Phi_T^\Sigma : \hat{\Sigma}^k \to \Sigma^k(T)$, $\Phi_T^V : \hat{V}^k \to V^k(T)$ and $\Phi_T^W : \hat{W}^k \to W^k(T)$ on element T by

$$\Phi_T^\Sigma(\hat{\tau}_h)(x) := \tau_h(x) := \tfrac{1}{J_T^2} F_T \hat{\tau}_h(\hat{x}) F_T^T \quad \text{for } \hat{\tau}_h \in \hat{\Sigma}^k, \tag{4.23}$$

$$\Phi_T^V(\hat{v}_h)(x) := v_h(x) := F_T^{-T} \hat{v}_h(\hat{x}) \quad \text{for } \hat{v}_h \in \hat{V}^k, \tag{4.24}$$

$$\Phi_T^W(\hat{w}_h)(x) := w_h(x) := \hat{w}_h(\hat{x}) \quad \text{for } \hat{w}_h \in \hat{W}^k. \tag{4.25}$$

Let now $\hat{v}_h \in \hat{V}^k$. By application of basic calculus, one can directly see, that the strain of $v_h := \Phi_T^V(\hat{v}_h)$ is given by

$$\varepsilon(v_h) = F_T^{-1} \hat{\varepsilon}(\hat{v}_h) F_T^{-T} + \sum_{i=1}^d \hat{v}_{h,i} F_T^{-1} (H_T^i)^{-1} F_T^{-T}.$$

On a mesh consisting of affine linear elements only, the Hessians H_T^i vanish, and we obtain $\varepsilon(v_h) = F_T^{-1}\hat{\varepsilon}(\hat{v}_h)F_T^{-T}$.

As one can easily check, these transformations preserve the respective degrees of freedom for the local spaces $W^k(T), V^k(T), \Sigma^k(T)$. We show these identities for the edge-based degrees of freedom for the displacements, as well as the facet-based nodal values for the stresses:

$$\int_E v_{h,\tau} q_i \, ds = \int_{\hat{E}} (\hat{v}_h^T F_T^{-1})(\frac{1}{J_E} F_T \hat{\tau}) \hat{q}_i J_E \, d\hat{s} = \int_{\hat{E}} \hat{v}_{h,\hat{\tau}} \hat{q}_i \, d\hat{s},$$

$$\int_F \sigma_{h,nn} q_i J_F \, ds = \int_{\hat{F}} (\frac{J_T}{J_F} F_T^{-T} \hat{n})^T (\frac{1}{J_T^2} F_T \hat{\sigma}_h F_T^T)(\frac{J_T}{J_F} F_T^{-T} \hat{n}) \hat{q}_i J_F^2 \, d\hat{s} = \int_{\hat{F}} \hat{\sigma}_{h,\hat{n}\hat{n}} \hat{q}_i \underbrace{J_{\hat{F}}}_{=1} \, d\hat{s}.$$

Here q_i are taken from the respective bases in the definitions of the degrees of freedom above. For all other degrees of freedom, similar identities can be shown along the same lines. All equivalences follow from basic calculus, but the respective terms become more lengthy.

These transformations can be used to map a nodal basis given explicitly on the reference element $\hat{T}$ to any element $T \in \mathcal{T}_h$, and thereby get a nodal basis there. Doing this, one ends up with the finite element spaces Σ_h^k, V_h^k, and W_h^k as defined in (4.13),(4.14).

4.2.2.6 A decomposition of the Nédélec space

The degrees of freedom for the finite element space V_h^k are associated to element edges, facets, or interiors. This enables us to split the space into subspaces corresponding to these quantities. Remember we called $\mathcal{X}_h = \mathcal{E}_h \cup \mathcal{F}_h \cup \mathcal{T}_h$ the union of all edges, facets, and elements in the mesh. The finite element space can be written as the direct sum

$$V_h^k = \bigoplus_{E \in \mathcal{E}_h} V_E^k \oplus \bigoplus_{F \in \mathcal{F}_h} V_F^k \oplus \bigoplus_{T \in \mathcal{T}_h} V_T^k = \bigoplus_{X \in \mathcal{X}_h} V_X^k. \tag{4.26}$$

Given a nodal basis for the Nédélec space, this splitting is induced by a splitting of this basis with respect to the different $X \in \mathcal{X}_h$. Note that all these subspaces are local, their support is restricted to only few elements.

Later on, it will prove useful not only to define the local spaces above, but also some global, low-order space. We call this space $V_{h,0}$. This space is the lowest-order Nédélec type I space, as introduced in [Néd80]. It consists of piecewise linear functions, where the tangential component along edges is constant. It corresponds to the lowest-order edge-based degrees of freedom above. We call the remaining part, which is built using the higher-order degrees of freedom, $\tilde{V}_h^k$. Similar to the splitting of V_h^k above, we can decompose $\tilde{V}_h^k$ into subspaces,

$$\tilde{V}_h^k = \bigoplus_{X \in \mathcal{X}_h} \tilde{V}_X^k.$$

In Section 4.4, an explicit basis for the Nédélec space will be given. Using this, we will then define all these spaces accurately.

4.3 Analysis of the TD-NNS method

In the sequel, we are concerned with the analysis of our mixed finite element method. We verify the conditions of Brezzi's theorem also in the discrete setting (Assumption 4.2). We introduce suitable nodal interpolation operators for the stress and displacement space, which lead to an a-priori bound for the discretization error. We provide a Korn-type inequality which relies on the splitting of V_h^k into a high- and a low-order part.

4.3.1 Stability properties

This section is devoted to verifying the discrete stability conditions from Assumption 4.2. On the continuous level, both $\Sigma = H(\operatorname{div}\operatorname{div})$ and $V = H(\operatorname{curl})$ are equipped with their natural norms. For the finite element analysis, we use a different set of norms. We use a broken H^1 norm for the displacements, and the L^2 norm for the stresses,

$$\|v_h\|_{V_h}^2 := \sum_{T \in \mathcal{T}_h} \|\varepsilon(v_h)\|_T^2 + \sum_{F \in \mathcal{F}_h} h_F^{-1} \|[v_h]_n\|_F^2, \tag{4.27}$$

$$\|\tau_h\|_{\Sigma_h} := \|\tau_h\|_{\Omega}. \tag{4.28}$$

Note that, using piecewise strains instead of piecewise gradients in $\|.\|_{V_h}$, we are able to avoid Korn's inequality. This will become useful when treating anisotropic geometries or elements, as arise in the discretization of shell- or beam-like structures. For a shape-regular mesh, the broken norm above is equivalent to one using piecewise gradients, which is frequently used in discontinuous Galerkin methods. It is then also possible not to take the full jump $[.]_n$, but its facet-wise projection $\Pi^1[.]_n$ onto the space of linear polynomials. We heavily use the theory provided in [Bre04].

Lemma 4.13. *Let $\mathcal{T}_h$ be a shape-regular, quasi-uniform triangulation of the domain Ω, then there exist constants $c_1, c_2 > 0$ depending only on Ω and the shape-regularity of $\mathcal{T}_h$ such that for all $v_h \in V_h^k$*

$$\|v_h\|_{V_h}^2 \geq c_1 \left(\sum_{T \in \mathcal{T}_h} \|\nabla v_h\|_T^2 + \sum_{F \in \mathcal{F}_h} h_F^{-1} \|[v_h]_n\|_F^2 \right), \tag{4.29}$$

$$\|v_h\|_{V_h}^2 \leq c_2 \left(\sum_{T \in \mathcal{T}_h} \|\varepsilon(v_h)\|_T^2 + \sum_{F \in \mathcal{F}_h} h_F^{-1} \|\Pi^1[v_h]_n\|_F^2 \right). \tag{4.30}$$

There Π^1 denotes the facet-wise projection onto $P^1(\mathcal{F}_h)$. These inequalities ensure equivalence of norms on V_h^k.

Proof. We first prove the lower bound (4.29). In [Bre04, Theorem 3.1], the existence of a constant c depending only on Ω and the shape-regularity of $\mathcal{T}_h$ was stated, such that

$$\sum_{T \in \mathcal{T}_h} \|\nabla v_h\|_T^2 \leq c \left(\sum_{T \in \mathcal{T}_h} \|\varepsilon(v_h)\|_T^2 + \Phi(v_h)^2 + \sum_{F \in \mathcal{F}_h} h_F^{-1} \|[v_h]_n\|_F^2 \right),$$

where Φ is a suitable semi-norm. Similar to [Bre04, equation (1.19)], we set

$$\Phi(v_h)^2 = \sum_{F \subset \Gamma} h_F^{-1} \|v_{h,n}\|_F^2.$$

Note that, in the original work, the whole of v_h was measured on the boundary, not only the normal component $v_{h,n}$. It is possible to use only this normal component due to the fact that $v_{h,\tau}$ vanishes on the non-trivial boundary part Γ_D. As this choice of Φ is absorbed by $\|.\|_{V_h}$, we directly obtain the desired result (4.29).

For the second inequality, we again apply theory from [Bre04]. Let now $D \subset \mathbb{R}^d$ be a bounded, connected Lipschitz domain. We call

$$RM(D) := \{v : D \to \mathbb{R}^d, \ x \mapsto a + \gamma x \,|\, a \in \mathbb{R}^d, \gamma \in \mathbb{R}_{SKW}\} \tag{4.31}$$

the space of (infinitesimal) rigid body motions. Let Π^{RM} denote the element-wise projection onto the space of piecewise rigid body motions $RM(\mathcal{T}_h)$. This projection can be defined element-wise via

$$\int_T v_h - \Pi^{RM} v_h \, dx = 0, \quad \int_T \mathrm{curl}(v_h - \Pi^{RM} v_h) \, dx = 0.$$

As $RM(T)$ is a subspace of $P^1(T)$, for any facet $F \in \mathcal{F}_h$,

$$\|(id - \Pi^1)[v_h]_n\|_F = \|(id - \Pi^1)[v_h - \Pi^{RM} v_h]_n\|_F \leq c \|[v_h - \Pi^{RM} v_h]_n\|_F.$$

Following [Bre04], the operator Π^{RM} satisfies the approximation property

$$\|[v_h - \Pi^{RM}(v_h)]_n\|_F^2 \leq c \sum_{T \in \Delta_F} h_F \|\varepsilon(v_h)\|_T^2 \quad \forall v_h \in V_h$$

where $c > 0$ is independent of the mesh size h. Combining these estimates, and using an argument of finite overlap, we obtain

$$\frac{1}{2} \sum_{F \in \mathcal{F}} h_F^{-1} \|[v_h]_n\|_F^2 \leq \sum_{F \in \mathcal{F}} \left[h_F^{-1} \|\Pi^1 [v_h]_n\|_F^2 + h_F^{-1} \|(id - \Pi^1)[v_h]_n\|_F^2 \right]$$

$$\leq c \left(\sum_{F \in \mathcal{F}} h_F^{-1} \|\Pi^1 [v_h]_n\|_F^2 + \sum_{T \in \mathcal{T}_h} \|\varepsilon(v_h)\|_T^2 \right).$$

This relation proves the second norm equivalence for the displacements. $\square$

Lemma 4.14. *On Σ_h, there holds the equivalence of norms*

$$\|\tau_h\|_{\Sigma_h}^2 \leq \|\tau_h\|_\Omega^2 + \sum_{F \in \mathcal{F}_h} h_F \|\tau_{h,nn}\|_F^2 \leq c_3 \|\tau_h\|_{\Sigma_h}^2,$$

given the triangulation $\mathcal{T}_h$ of Ω is shape regular.

Proof. The lower bound is trivial, we prove the upper bound by a scaling argument. We transform the domain of integration element-wise to the reference element, using the $H(\text{div div})$ conforming transformation Φ_T^Σ given in (4.23). We do all calculations in detail here; we will often use a similar approach later in this work, but then abandon this level of exactness. Let now $\tau_h \in \Sigma_h$ be arbitrary, the shape-regularity of $\mathcal{T}_h$ ensures that $|F_T|_s \simeq h_T$ on each element T, and therefore

$$\|\tau_h\|_T^2 = \int_{\hat{T}} \frac{1}{J_T^4} (F_T \hat{\tau}_h F_T^T)^2 J_T \, d\hat{x} \simeq h_T^{4-3d} \int_{\hat{T}} |\hat{\tau}_h|^2 d\hat{x}.$$

On the finite dimensional space $\hat{\Sigma}^k$ on the reference element, there holds

$$\int_{\hat{T}} |\hat{\tau}_h|^2 d\hat{x} \geq c \sum_{\hat{F} \in \mathcal{F}(\hat{T})} \int_{\hat{F}} |\hat{\tau}_{h,\hat{n}\hat{n}}|^2 d\hat{s}$$

Thereby we conduct

$$\begin{aligned}
\|\tau_h\|_T^2 &\geq ch_T^{4-3d} \sum_{\hat{F} \in \mathcal{F}(\hat{T})} \int_{\hat{F}} \hat{\tau}_{h,\hat{n}\hat{n}}^2 d\hat{s} \\
&= ch_T^{4-3d} \sum_{F \in \mathcal{F}(T)} \int_F |(\hat{n}^T F_T^{-1})(F_T \hat{\tau}_h F_T^T)(F_T^{-T} \hat{n})|^2 \frac{1}{J_F} ds \\
&= ch_T^{4-3d} \sum_{F \in \mathcal{F}(T)} \int_F \frac{J_T^4 J_F^4}{J_T^4} |\tau_{h,nn}|^2 \frac{1}{J_F} ds \\
&\simeq \sum_{F \in \mathcal{F}(T)} h_F \|\tau_h\|_F^2.
\end{aligned}$$

Here we used that, on a shape-regular triangulation, we have $h_T \simeq h_F$ for all $T \in \mathcal{T}_h$, $F \in \mathcal{F}(T)$. This estimate concludes the proof of the lemma. $\square$

The next three lemmas state that the finite element spaces Σ_h^k, V_h^k are a stable pair for discretizing the equations of elasticity.

Lemma 4.15. *The bilinear forms $a(\cdot,\cdot)$, $b(\cdot,\cdot)$ are bounded on Σ_h^k, V_h^k in the respective discrete norms, there exist constants $\tilde{c}_{a,2}, \tilde{c}_{b,2} > 0$ such that*

$$a(\sigma_h, \tau_h) \leq \tilde{c}_{a,2} \|\sigma_h\|_{\Sigma_h} \|\tau_h\|_{\Sigma_h} \qquad \forall \sigma_h, \tau_h \in \Sigma_h^k,$$
$$b(\tau_h, v_h) \leq \tilde{c}_{b,2} \|\tau_h\|_{\Sigma_h} \|v_h\|_{V_h} \qquad \forall \tau_h \in \Sigma_h^k, v_h \in V_h^k.$$

The constants $\tilde{c}_{a,2}, \tilde{c}_{b,2}$ are independent of the mesh size h.

Proof. It is straightforward to prove continuity of $a(\cdot,\cdot)$ in the discrete norm $\|.\|_{\Sigma_h}$. For the constant of boundedness we obtain

$$\tilde{c}_{a,2} = \lambda_{max}(\bar{A}) = \frac{1}{2\bar{\mu}},$$

where $\lambda_{max}(\bar{A})$ is the maximal eigenvalue of the symmetric fourth order tensor $\bar{A}$. The last identity holds for isotropic, linear elastic, homogenous materials. One can see that this bound does not deteriorate for nearly incompressible materials.

Continuity of $b(\cdot,\cdot)$ also follows directly when using representation (4.19) and the norm equivalence on the discrete level stated in Lemma 4.14. □

We have seen that, in the continuous setting, it is possible to show coercivity of $a(\cdot,\cdot)$ on the kernel of B; the constant of coercivity is then independent of the Lamé parameter $\bar{\lambda}$. In the discrete setting, we cannot provide such a result for $a(\cdot,\cdot)$. In Chapter 5, we propose a stabilized bilinear form which is coercive independently of $\bar{\lambda}$.

Lemma 4.16. *The bilinear form $a(\cdot,\cdot)$ is coercive on the space Σ_h^k; the constant of coercivity $\tilde{c}_{a,1} > 0$, which satisfies*

$$a(\tau_h, \tau_h) \geq \tilde{c}_{a,1} \|\tau_h\|_{\Sigma_h} \qquad \forall \tau_h \in \Sigma_h^k$$

is independent of the mesh size h, but depends on the Lamé parameter $\bar{\lambda}$ as $\tilde{c}_{a,1}(\bar{\lambda}) = \mathcal{O}(1/\bar{\lambda})$.

Proof. The lemma follows directly from the fact that

$$\tilde{c}_{a,1} = \lambda_{min}(\bar{A}) = \frac{1}{3\lambda + 2\mu}.$$

□

Our next aim is proving an inf-sup condition for $b(\cdot,\cdot)$. Let therefore $\hat{T}$ be the reference element, with (simplicial) facets $\hat{F}_i$, $i = 1, \ldots, d+1$. In Section 4.4, we provide *facet basis tensors* $\hat{S}^{F_i}$, $i = 1, \ldots, d+1$. These tensors are constant on the element, and their normal-normal component vanishes on all facets but $\hat{F}_i$. In two space dimensions, the family $\{\hat{S}^{F_i} : i = 1, 2, 3\}$ is a basis for $\hat{\Sigma}^0 = P^0_{SYM}(\hat{T})$. Conversely, in three dimensions, the four constant tensor fields $\hat{S}^{F_i}$ cannot span the six-dimensional space $P^0_{SYM}(\hat{T})$. There exist two further, linearly independent fields $\hat{S}^{T,1}, \hat{S}^{T,2}$, which can be constructed such that their normal-normal component vanishes on the whole boundary. Then the family $\{\hat{S}^{F_i} : i = 1, \ldots, 4\} \cup \{\hat{S}^{T,1}, \hat{S}^{T,2}\}$ provides a proper basis for $\hat{\Sigma}^0 = P^0_{SYM}(\hat{T})$. Using the conforming transformation Φ^Σ_T, one can build the global space Σ_h^0.

In the following, we will additionally need facet-based finite element subspaces Σ_F^k, which have support only on the two elements neighboring the facet F. Moreover, their normal-normal component restricted to F shall span $P^k(F)$, and vanish on all other facets,

$$\{\sigma_{h,nn}^F|_F : \sigma_h^F \in \Sigma_F^k\} = P^k(F),$$
$$\sigma_{h,nn}^F|_{\tilde{F}} = 0 \quad \text{for } \sigma_h^F \in \Sigma_F^k, \ \tilde{F} \in \mathcal{F}_h, \ \tilde{F} \neq F.$$

This means that the space Σ_F^k corresponds to the degrees of freedom of facet F. A suitable basis for this space will be provided in Section 4.4. We note that these spaces are linearly independent.

Let now λ_i be the barycentric coordinate for the vertex opposite facet F_i. Then, $\lambda_i = 0$ on F_i, the tensor fields

$$\hat{B}^i := \lambda_i \hat{S}^{F_i}, \quad i = 1, \ldots, d+1$$

are bubble functions, i.e. their normal-normal component vanishes on the whole boundary $\partial \hat{T}$. In 3D, we additionally set

$$\hat{B}^5 = \hat{S}^{T,1}, \quad \hat{B}^6 = \hat{S}^{T,2},$$

as these fields are already element bubbles. We call

$$\hat{\Sigma}_T^k := \operatorname{span}\{\hat{B}^i\}$$

the *bubble space* on the reference element. Again, Σ_T^k shall denote its transformation to an element $T \in \mathcal{T}_h$.

Lemma 4.17. *Let $\mathcal{T}_h$ be a quasi-uniform shape-regular triangulation of Ω. There holds the stability estimate*

$$\inf_{v_h \in V_h^k} \sup_{\tau_h \in \Sigma_h^k} \frac{b(\tau_h, v_h)}{\|\tau_h\|_{\Sigma_h} \|v_h\|_{V_h}} \geq \tilde{c}_{b,1},$$

where $\tilde{c}_{b,1} > 0$ is independent of the mesh size h.

Proof. The finite element space Σ_h^k can be decomposed in two parts,

$$\Sigma_h^k = \Sigma_h^{k,f} \oplus \Sigma_h^{k,b}.$$

Here $\Sigma_h^{k,f}$ is associated to the degrees of freedom lying on element facets, while $\Sigma_h^{k,b}$ is called "bubble space" and consists of element bubble functions which correspond to the degrees of freedom interior to one element.

The facet space $\Sigma_h^{k,f}$ is the direct sum of contributions coming from the different facets $F \in \mathcal{F}_h$,

$$\Sigma_h^{k,f} = \bigoplus_{F \in \mathcal{F}_h} \Sigma_F^k.$$

The bubble space $\Sigma_h^{k,b}$ is built from element bubble spaces Σ_T^k described above,

$$\Sigma_h^{k,b} = \bigoplus_{T \in \mathcal{T}_h} \Sigma_T^k.$$

Note that the two spaces are linearly independent and the finite element functions have bounded overlap. Thus, there exists a constant independent of h such that

$$\|\tau_h^f\|_{\Sigma_h} + \|\tau_h^b\|_{\Sigma_h} \leq c\|\tau_h\|_{\Sigma_h} \quad \text{for all } \tau_h = \tau_h^f + \tau_h^b, \; \tau_h^f \in \Sigma_h^{k,f}, \; \tau_h^b \in \Sigma_h^{k,b}.$$

Let now $v_h \in V_h^k$ be given. We construct $\tau_h = c_f \tau_h^f + c_b \tau_h^b$, where $\tau_h^f \in \Sigma_h^{k,f}$, $\tau_h^b \in \Sigma_h^{k,b}$, and $c_f, c_b \in \mathbb{R}$ are to be specified below. Let τ_h^f be such that

$$\tau_{h,nn}^f|_F = h_F^{-1} [\![v_h]\!]_{n,F}.$$

This is possible, as $[\![v_h]\!]_{n,F}$ is of polynomial degree k for each facet F.

Next, we construct the bubble part τ_h^b. We use the bubble functions $\hat{B}^i$ defined on the reference element. Each of these bubbles is linked to some $\hat{S}^{F_i}, \hat{S}^{T,j}$ by definition. For simplicity of notation, we do the following calculations for the case $d = 2$, as then we have $\hat{B}^i = \lambda_i \hat{S}^{F_i}, i = 1, 2, 3$. For $d = 3$, all estimates work along the same lines, using both $\hat{S}^{F_i}, \hat{S}^{T,j}$. We define τ_h^b element-wise via its representation $\hat{\tau}_h^b$ on the reference element:

$$\hat{\tau}_h^b := \frac{J_T^2}{h_T^4} \sum_{i=1}^3 \underbrace{\left(\hat{\varepsilon}(\hat{v}_h) : \hat{S}^{F_i}\right)}_{\in P^{k-1}(T)} \underbrace{\hat{B}^i}_{\in P^1(T)}.$$

We now estimate the two parts separately. As $\Sigma_h^{k,f}$ consists of facet-bound fields only, we obtain similar to Lemma 4.14,

$$\|\tau_h^f\|_{\Sigma_h}^2 \leq c_1 \sum_{F \in \mathcal{F}_h} h_F \|\tau_{h,nn}^f\|_F^2 = c_1 \sum_{F \in \mathcal{F}_h} h_F^{-1} \|[\![v_h]\!]_n\|_F^2.$$

For the bubble part, we can show the following bound by transformation to the reference element:

$$\|\tau_h^b\|_{\Sigma_h}^2 \leq c_2^2 \sum_{T \in \mathcal{T}_h} \|\varepsilon(v_h)\|_T^2.$$

Thus, we may deduce

$$\|\tau_h\|_{\Sigma_h} \leq \max(c_1 c_f, c_2 c_b) \|v_h\|_{V_h}. \tag{4.32}$$

Next, we show the existence of $c_3 > 0$ such that, on each element $T \in \mathcal{T}_h$,

$$\int_T \tau_h^b : \varepsilon(v_h) \, dx \geq c_3 \|\varepsilon(v_h)\|_T^2.$$

We do this again by transformation to the reference element, employing that by shape-regularity $|F_T^{-1}|_s \simeq h_T^{-1}$.

$$
\begin{aligned}
\int_T \tau_h^b : \varepsilon(v_h)\, dx &= \int_{\hat T} \frac{1}{J_T^2} \hat\tau_h^b : \hat\varepsilon(\hat v_h) J_T\, d\hat x \\
&= \int_{\hat T} h_T^{-4} \sum_{i=1}^3 \left(\hat\varepsilon(\hat v_h) : \hat S^{F_i}\right) \hat B^i : \hat\varepsilon(\hat v_h) J_T\, d\hat x \\
&= \int_{\hat T} h_T^{-4} \sum_{i=1}^3 \left(\hat\varepsilon(\hat v_h) : \hat S^{F_i}\right)^2 \lambda_i J_T\, dx \\
&\simeq \int_{\hat T} h_T^{-4} |\hat\varepsilon(\hat v_h)|^2 J_T\, d\hat x \\
&\simeq \int_{\hat T} |F_T^{-1} \hat\varepsilon(\hat v_h) F_T^{-T}|^2 J_T\, d\hat x = \|\varepsilon(v_h)\|_T^2.
\end{aligned}
$$

We used that the $\hat B^i$ are linearly independent, and that the $\hat S^{F_i}$ form a basis for the piecewise constant, symmetric tensor fields in 2D. We can now show the following lower bound for $b(\tau_h, v)$, where we use the estimates from above, as well as Young's inequality in the last line:

$$
\begin{aligned}
b(\tau_h, v_h) &= \sum_{T \in \mathcal T_h} \int_T \varepsilon(v_h) : \tau_h\, dx - \sum_{F \in \mathcal F_h} \int_F \tau_{h,nn} [\![v_h]\!]_n\, ds \\
&= \sum_{T \in \mathcal T_h} \int_T \varepsilon(v_h) : (c_f \tau_h^f + c_b \tau_h^b)\, dx - \sum_{F \in \mathcal F_h} \int_F c_f \tau_{h,nn}^f [\![v_h]\!]_n\, ds \\
&\geq \sum_{T \in \mathcal T_h} \left[c_b c_3 \|\varepsilon(v_h)\|_T^2 - c_f \|\tau_h^f\|_T \|\varepsilon(v_h)\|_T\right] + \sum_{F \in \mathcal F_h} c_f h_F^{-1} \|[\![v_h]\!]_n\|_F^2 \\
&\geq \sum_{T \in \mathcal T_h} \left[c_b c_3 \|\varepsilon(v_h)\|_T^2 - \sum_{F \subset \partial T} c_f c_1 h_F^{-1/2} \|[\![v_h]\!]_n\|_F \|\varepsilon(v_h)\|_T\right] + \sum_{F \in \mathcal F_h} c_f h_F^{-1} \|[\![v_h]\!]_n\|_F^2 \\
&\geq \sum_{T \in \mathcal T_h} \left(c_b c_3 - \frac{c_f c_1 \gamma^2}{2}\right) \|\varepsilon(v_h)\|_T^2 + \sum_{F \in \mathcal F_h} c_f h_F^{-1} \left(1 - \frac{c_1}{2\gamma^2}\right) \|[\![v_h]\!]_n\|_F^2.
\end{aligned}
$$

Setting $\gamma^2 = c_1, c_f = 1, c_b = (1 + c_1^2)/(2c_3)$, the estimate above together with (4.32) yields the required result,

$$
b(\tau_h, v_h) \geq \frac{1}{2} \|v\|_{V_h}^2 \geq \min\left(\frac{1}{2c_1}, \frac{c_3}{c_2(1 + c_1^2)}\right) \|\tau_h\|_{\Sigma_h} \|v\|_{V_h}.
$$

□

Corollary 4.18. *Let $\mathcal T_h$ be a shape-regular triangulation of Ω, and define*

$$\Sigma_h^{k, k_f} := \{\tau_h \in \Sigma_h^k : \tau_{h,nn}|_F \in P^{k_f}(F)\ \forall F \in \mathcal F_h\}.$$

Then the statement of Lemma 4.17 still holds on $\Sigma_h^{k,1} \times V_h^k$.

Proof. We obtain this statement when choosing τ_h^f in the proof of Lemma 4.17 like

$$\tau_{h,nn}^f|_F = h_F^{-1}\Pi^1[\![v]\!]_{n,F}.$$

Then we get stability using the norm equivalence (4.30) in Lemma 4.13. □

4.3.2 Nodal interpolation and error estimates

This section is devoted to giving a-priori error estimates for the difference between the exact solution (σ, u) and the Galerkin approximation (σ_h, u_h). Lemma 4.3 states, that the error can be estimated by the best-approximation error of $\Sigma_h^k \times V_h^k$. We bound this quantity by the interpolation error for suitable interpolators for the finite element spaces. We propose to use nodal interpolation operators as implied by the finite elements for Σ_h^k, V_h^k. First we shortly present the nodal interpolation operator $\mathcal{I}_{h,k}^W$ for the space W_h^k, and recall its approximation properties. For the displacement space V_h^k, the nodal interpolation operator $\mathcal{I}_{h,k}^V$ is the well-known Nédélec interpolation operator, as defined e.g. in [Mon03]. For the stress space, we introduce and analyze the corresponding nodal interpolation operator $\mathcal{I}_{h,k}^\Sigma$. Last, we observe that the nodal interpolation operator $\mathcal{I}_{h,k}^\mathcal{P}$ for the L^2 conforming space $\mathcal{P}_h^k$ reduces to an element-wise L^2 projection.

One major drawback of the nodal interpolation operators is that they are usually not defined on the whole space of interest, e.g. on H^1, $H(\operatorname{curl}), H(\operatorname{div}\operatorname{div})$ or L^2. This difficulty can be circumvented by the usage of quasi-interpolation operators, which require less smoothness, and are well defined on $L^2(\Omega)$. Most famous among them is probably the Clément quasi-interpolation operator introduced in [Clé75], which was constructed for continuous finite elements. Generalizations of this interpolator are e.g. the Scott-Zhang interpolation operator [SZ80], which preserves boundary conditions, or the family of quasi-interpolation operators for H^1, $H(\operatorname{curl})$, $H(\operatorname{div})$ and L^2, which was introduced in [Sch01]. The operators stemming from the latter family satisfy a commuting diagram property, and can be constructed such that degrees of freedom are preserved for polynomial functions. We do a careful analysis in Section 6.2.1. Last, we refer to the family of projection-based interpolation operators, which also can be defined on H^1, $H(\operatorname{curl})$, $H(\operatorname{div})$ and L^2, and commute with differential operators [DB05].

4.3.2.1 An interpolation operator for H^1

We shortly recall the definition of the nodal interpolation operator $\mathcal{I}_{h,k}^W$ for H^1. It is defined using the degrees of freedom of the high-order H^1 finite element:

$$N^W(w - \mathcal{I}_{h,k}^W w) = 0 \qquad \forall N^W \in \mathcal{N}_k^W. \tag{4.33}$$

The following theorem provides the standard result on the interpolation error for the nodal operator. A proof can be found e.g. in [BS02, Mon03].

Theorem 4.19. *Let $\mathcal{T}_h$ be a shape-regular, uniform triangulation of Ω. Let $\mathcal{I}_{h,k}^W$ denote the nodal interpolant for W_h^k as defined in (4.33). Let $s \in [d/2 + \delta, k + 1]$, $\delta > 0$ such that $w \in H^s(\Omega)$, then*

$$\|w - \mathcal{I}_{h,k}^W w\|_{H^1(\Omega)} \leq c h^{s-1} |w|_{H^s(\Omega)}.$$

The restriction $w \in H^s(\Omega)$ for $s > d/2 + \delta$ ensures that all nodal values can be evaluated, and the interpolation operator $\mathcal{I}_{h,k}^W$ is well defined. If this is not the case, one can obtain a similar estimate using one of the quasi-interpolation operators mentioned above.

4.3.2.2 An interpolation operator for the Nédélec space

The nodal interpolation operator for the Nédélec space V_h^k is defined via the degrees of freedom for the Nédélec element:

$$N^V(v - \mathcal{I}_{h,k}^V v) = 0 \qquad \forall N^V \in \mathcal{N}_k^V. \tag{4.34}$$

The following theorem concerning the approximation properties of this interpolator is widely used for error estimates. A proof can be found in [Mon03].

Theorem 4.20. *Let $\mathcal{T}_h$ be a shape-regular, quasi-uniform triangulation of Ω. Let $\mathcal{I}_{h,k}^V$ denote the nodal interpolant for V_h^k defined as above. If $v \in H^s(\mathrm{curl}; \Omega)$ for some $s \in [d/2 - 1 + \delta, k]$ for $\delta > 0$, then*

$$\|v - \mathcal{I}_{h,k}^V v\|_{H(\mathrm{curl})} \leq c h^s \left(\|v\|_{H^s(\Omega)} + \|\mathrm{curl}\, v\|_{H^s(\Omega)} \right).$$

The restriction $v \in H^s(\mathrm{curl}; \Omega)$, $s > d/2 - 1 + \delta$ in the theorem above ensures that all nodal values are well-defined for v. Otherwise, one has to use quasi-interpolation operators as mentioned before. For the remainder of this section, we will use the nodal Nédélec interpolation operator $\mathcal{I}_{h,k}^V$ on $H(\mathrm{curl})$. Therefore, we assume that the solution u to the elasticity problem is sufficiently smooth, such that the interpolation operator is well defined. Otherwise, all estimates can be done using the local averaging operators described above.

For error analysis in the TD-NNS method, we need to quantify the approximation properties of the Nédélec interpolation operator in the discrete norm $\|\cdot\|_{V_h}$.

Theorem 4.21. *Let $\mathcal{T}_h$ be a regular triangulation of Ω. Let $\mathcal{I}_{h,k}^V$ denote the nodal interpolant for V_h^k defined as above. Let $v \in H^s(\mathrm{curl}; \Omega)$ with $s \in [d/2 - 1 + \delta, k]$ for $\delta > 0$ satisfy $v|_T \in H^{k+1}(T)$ for all elements $T \in \mathcal{T}_h$. Then there holds for $1 \leq m \leq k$*

$$\|v - \mathcal{I}_{h,k}^V v\|_{V_h} \leq c \left(\sum_{T \in \mathcal{T}_h} h_T^{2m} \|\nabla^m \varepsilon(v)\|_T^2 \right)^{1/2}.$$

The generic constant c depends on the shape-regularity of the triangulation, but not on the mesh size h.

Proof. Let v fulfilling the conditions above be arbitrary, and let $T \in \mathcal{T}_h$. Due to [Mon03, Lemma 5.38] and the Sobolev embedding theorem (see e.g. [GR86, Wer02]), we have that the local interpolation operator $\mathcal{I}_k^{\hat{V}}$ on the reference element is continuous on $H^s(\mathrm{curl};\hat{T})$. From this, we deduce that also $\mathcal{I}_k^{\hat{V}} : H^{m+1}(\hat{T}) \to H^1(\hat{T})$ is bounded; let $\|\mathcal{I}_k^{\hat{V}}\|$ be the corresponding operator norm.

We now use a similar approach as in [BS02, Theorem 4.4.4]. We use that $\mathcal{I}_k^{\hat{V}}$ preserves polynomials up to order k, and estimate

$$\begin{aligned}
\|\hat{v} - \mathcal{I}_k^{\hat{V}}\hat{v}\|_{H^1(\hat{T})} &\leq \inf_{\hat{q} \in [P^k(\hat{T})]^d} \|\hat{v} - \hat{q}\|_{H^1(\hat{T})} + \|\mathcal{I}_k^{\hat{V}}(\hat{q} - \hat{v})\|_{H^1(\hat{T})} \\
&\leq \inf_{\hat{q} \in [P^k(\hat{T})]^d} \left(1 + \|\mathcal{I}_k^{\hat{V}}\|\right) \|\hat{v} - \hat{q}\|_{H^{m+1}(\hat{T})} \\
&\leq c\left(1 + \|\mathcal{I}_k^{\hat{V}}\|\right) |\hat{v}|_{H^{m+1}(\hat{T})}.
\end{aligned}$$

There, we used the Lemma of Bramble-Hilbert (Lemma 4.3.8 in [BS02]) in the last line.

Using this local estimate on the reference element, we proceed

$$\begin{aligned}
\|v - \mathcal{I}_{h,k}^V v\|_{V_h}^2 &= \sum_{T \in \mathcal{T}_h} \|\varepsilon(v - \mathcal{I}_k^{V(T)}v)\|_T^2 + \sum_{F \in \mathcal{F}_h} h_F^{-1} \|[v - \mathcal{I}_k^{V(\Delta_F)}v]_n\|_F^2 \\
&\leq \sum_{T \in \mathcal{T}_h} \left[\|\varepsilon(v - \mathcal{I}_k^{V(T)}v)\|_T^2 + h_T^{-1}\|v - \mathcal{I}_k^{V(T)}v\|_{\partial T}^2\right].
\end{aligned}$$

Here we reordered the facet integrals element-wise in the last line, and estimated the jump in normal direction by the full trace. For an element $T \in \mathcal{T}_h$, we transform the respective summand to the reference element $\hat{T}$

$$\begin{aligned}
\|\varepsilon(v - \mathcal{I}_k^{V(T)}v)\|_T^2 + h_T^{-1}\|v_n - \mathcal{I}_k^{V(T)}v_n\|_{\partial T}^2 &= \\
= \int_{\hat{T}} |F_T^{-1}\hat{\varepsilon}(\hat{v} - \mathcal{I}_k^{\hat{V}}\hat{v})F_T^{-T}|^2 J_T \, d\hat{x} &+ \int_{\partial \hat{T}} h_T^{-1}|(\hat{v} - \mathcal{I}_k^{\hat{V}}\hat{v})F_T^{-1}|^2 J_F \, d\hat{s} \\
&\leq ch_T^{-4+d}\left(\|\hat{\varepsilon}(\hat{v} - \mathcal{I}_k^{\hat{V}}\hat{v})\|_{\hat{T}}^2 + \|\hat{v} - \mathcal{I}_k^{\hat{V}}\hat{v}\|_{\partial \hat{T}}^2\right) \\
&\leq ch_T^{-4+d}\|\hat{v} - \mathcal{I}_k^{\hat{V}}\hat{v}\|_{H^1(\hat{T})}^2 \\
&\leq ch_T^{-4+d}|\hat{v}|_{H^{m+1}(\hat{T})}^2.
\end{aligned}$$

Here we used the local approximation estimate on the reference element $\hat{T}$. Transformation back to element T yields again

$$\|\varepsilon(v - \mathcal{I}_k^{V(T)}v)\|_T^2 + h_T^{-1}\|v_n - \mathcal{I}_k^{V(T)}v_n\|_{\partial T}^2 \leq ch_T^{2m}|v|_{H^{m+1}(T)}^2.$$

Summarizing, we obtain the required result

$$\|v - \mathcal{I}_{h,k}^V v\|_{V_h} \leq c \Big(\sum_{T \in \mathcal{T}_h} h_T^{2m} |v|_{H^{m+1}(T)}^2 \Big)^{1/2} \leq c \Big(\sum_{T \in \mathcal{T}_h} h_T^{2m} |\varepsilon(v)|_{H^m(T)}^2 \Big)^{1/2}.$$

In the last step, we used that for integer $m \geq 1$ one can show $|\varepsilon(v)|_{H^m(T)} \leq c|v|_{H^{m+1}(T)}$ by a direct evaluation of the respective differential terms. Thus, Korn's inequality is not needed in this step. □

4.3.2.3 An interpolation operator for the stress space

As for the Nédélec space, we use the nodal interpolation operator $\mathcal{I}_{h,k}^\Sigma$ implied by the degrees of freedom of the normal-normal continuous finite element:

$$N^\Sigma(\tau - \mathcal{I}_{h,k}^\Sigma \tau) = 0 \quad \forall N^\Sigma \in \mathcal{N}_k^\Sigma. \tag{4.35}$$

Again, this operator is not well defined on $L^2(\Omega)$, but on $\Sigma = H(\text{div div})$: Due to the trace theorem, we have that the normal-normal trace of some element $\tau \in \Sigma$ can be tested against polynomial functions, thus the facet-bound degrees of freedom are well-defined. The interior degrees of freedom can be evaluated for any $\tau \in L^2(\Omega)$. As the local space spans $P_{SYM}^k(T)$ on each element T, we expect convergence of order $k+1$ in the L^2 norm. We first show stability of the operator.

Lemma 4.22. *Let $\mathcal{T}_h$ be a regular triangulation of Ω. Let $\tau \in L^2(\Omega)$ be such that $\tau_{nn}|_F \in L^2(F)$ for all facets $F \in \mathcal{F}_h$. Then the nodal interpolant $\mathcal{I}_{h,k}^\Sigma$ for Σ_h^k defined by equation (4.35) is bounded, its norm $\|\mathcal{I}_{h,k}^\Sigma\|$ satisfies*

$$\|\mathcal{I}_{h,k}^\Sigma \tau\|_{\Sigma_h} \leq \|\mathcal{I}_{h,k}^\Sigma\| \Big(\|\tau\|_\Omega^2 + \sum_{F \in \mathcal{F}_h} h_F \|\tau_{nn}\|_F^2 \Big)^{1/2}.$$

Here, $\|\mathcal{I}_{h,k}^\Sigma\|$ does not depend on the mesh size h.

Proof. We show the estimate on the reference element, i.e.

$$\|\mathcal{I}_k^{\hat{\Sigma}} \hat{\tau}\|_{\hat{T}}^2 \leq c \left(\|\hat{\tau}\|_{\hat{T}}^2 + \|\hat{\tau}_{\hat{n}\hat{n}}\|_{\partial \hat{T}}^2 \right).$$

Let $\{\hat{\varphi}_{\hat{F}_i,j}^\Sigma, \hat{\varphi}_{\hat{T},l}^\Sigma : j \in I_{\hat{F}_i}, l \in I_{\hat{T}}\}$ be a nodal basis for the stress finite element with degrees of freedom $\mathcal{N}_k^\Sigma(\hat{T}) = \{N_{\hat{F}_i,j}^\Sigma, N_{\hat{T},l}^\Sigma\}$, as they were defined in (4.20), (4.21). Then we have

$$\begin{aligned}
\|\mathcal{I}_k^\Sigma \hat{\tau}\|_{\hat{T}} &= \left\| \sum_{\hat{F}_i \in \mathcal{F}(\hat{T})} \sum_{j \in I_{\hat{F}_i}} (N_{\hat{F}_i,j}^\Sigma \hat{\tau}) \hat{\varphi}_{\hat{F}_i,j}^\Sigma + \sum_{l \in I_{\hat{T}}} (N_{\hat{T},l}^\Sigma \hat{\tau}) \hat{\varphi}_{\hat{T},l}^\Sigma \right\|_{\hat{T}} \\
&\leq \sum_{\hat{F}_i \in \mathcal{F}(\hat{T})} \sum_{j \in I_{\hat{F}_i}} |N_{\hat{F}_i,j}^\Sigma \hat{\tau}| \|\hat{\varphi}_{\hat{F}_i,j}^\Sigma\|_{\hat{T}} + \sum_{l \in I_{\hat{T}}} |N_{\hat{T},l}^\Sigma \hat{\tau}| \|\hat{\varphi}_{\hat{T},l}^\Sigma\|_{\hat{T}} \\
&\leq c \left[\sum_{\hat{F}_i \in \mathcal{F}(\hat{T})} \sum_{j \in I_{\hat{F}_i}} |N_{\hat{F}_i,j}^\Sigma \hat{\tau}| + \sum_{l \in I_{\hat{T}}} |N_{\hat{T},l}^\Sigma \hat{\tau}| \right].
\end{aligned}$$

In the last line we used that the nodal basis on the reference element is finite and therefore is bounded uniformly from above and below in the L^2 norm. It remains to bound the absolute values of the degrees of freedom. For the facet-based nodal values, we obtain, using the basis $\{q_j : j \in I_{\hat{F}_i}\}$ of $P^k(\hat{F}_i)$ taken from (4.20),

$$\begin{aligned}
|N_{\hat{F}_i,j}^\Sigma \hat{\tau}| &= \left| \int_{\hat{F}_i} \hat{\tau}_{\hat{n}\hat{n}} q_j J_{\hat{F}_i} \, ds \right| = \left| \int_{\hat{F}_i} \hat{\tau}_{\hat{n}\hat{n}} q_j \, ds \right| \\
&\leq \|\hat{\tau}_{\hat{n}\hat{n}}\|_{\hat{F}_i} \|q_j\|_{\hat{F}_i} \leq c \|\hat{\tau}_{\hat{n}\hat{n}}\|_{\hat{F}_i}.
\end{aligned}$$

Similarly, we see for the interior degrees of freedom that

$$\begin{aligned}
|N_{\hat{T},l}^\Sigma \hat{\tau}| &= \left| \int_{\hat{T}} J_{\hat{T}} \hat{\tau} : F_{\hat{T}}^{-T} \gamma_l F_{\hat{T}}^{-1} \, dx \right| = \left| \int_{\hat{T}} \hat{\tau} : \gamma_l \, dx \right| \\
&\leq \|\hat{\tau}\|_{\hat{T}} \|\gamma_l\|_{\hat{T}} \leq c \|\hat{\tau}\|_{\hat{T}},
\end{aligned}$$

where $\{\gamma_l : l \in I_{\hat{T}}\}$ is the basis of $P_{0,nn}^k(\hat{T})$ used in (4.21). A scaling argument using the transformation $\Phi_{\hat{T}}^\Sigma$ directly leads to the h-dependent estimate from the Lemma. □

As the local space $\Sigma^k(T)$ is a full polynomial space for $T \in \mathcal{T}_h$, we are able to show an optimal order of approximation for the nodal interpolation operator.

Theorem 4.23. *Let $\mathcal{T}_h$ be a shape-regular triangulation of Ω, and let $\mathcal{I}_{h,k}^\Sigma$ be the nodal interpolation operator defined in (4.35). Let $1 \leq m \leq k+1$, and $\tau \in L^2(\Omega)$ such that $\tau|_T \in H^m(T)$ for all elements $T \in \mathcal{T}_h$. Then the interpolation error is bounded by*

$$\|\tau - \mathcal{I}_{h,k}^\Sigma \tau\|_{\Sigma_h} \leq c \left(\sum_{T \in \mathcal{T}_h} h_T^{2m} |\tau|_{H^m(T)}^2 \right)^{1/2}.$$

The constant c depends on the shape-regularity of the triangulation, but not on the mesh size h.

Proof. The proof for this theorem runs very much along the same lines as the one for [BS02, Theorem 4.4.4] or for Theorem 4.21. Again, we can verify, using that $\hat{\Sigma}^k$ spans $P_{SYM}^k(\hat{T})$, that on the reference element

$$\|\hat{\tau} - \mathcal{I}_k^{\hat{\Sigma}}\hat{\tau}\|_{\hat{T}} \leq c \inf_{\hat{\gamma} \in P_{SYM}^k(\hat{T})} \left(\|\hat{\tau} - \hat{\gamma}\|_{\hat{T}} + \|\hat{\tau}_{\hat{n}\hat{n}} - \hat{\gamma}_{\hat{n}\hat{n}}\|_{\partial \hat{T}}\right)$$
$$\leq c \inf_{\hat{\gamma} \in P_{SYM}^k(\hat{T})} \|\hat{\tau} - \hat{\gamma}\|_{H^m(\hat{T})} \leq c|\hat{\tau}|_{H^m(\hat{T})}.$$

Transformation to an arbitrary element $T \in \mathcal{T}_h$ using the conforming transformation Φ_T^Σ yields then the scaled estimate. □

4.3.2.4 An interpolation operator for the space $\mathcal{P}_h^k$

Last, we define the nodal interpolation operator for the L^2 conforming space $\mathcal{P}_h^k$. We already mentioned in Section 4.2.2.4 that the nodal interpolation operator coincides with the element-wise L^2 projection onto the local finite element space. Therefore, the following interpolation error estimate follows directly.

Lemma 4.24. *Let $\mathcal{T}_h$ be a shape-regular triangulation of Ω, and let $\mathcal{I}_{h,k}^\mathcal{P}$ be the nodal interpolation operator for $\mathcal{P}_h^k$. Let $1 \leq m \leq k+1$, and $p \in H^m(\Omega)$. Then the interpolation error is bounded by*

$$\|p - \mathcal{I}_{h,k}^\mathcal{P} p\|_\Omega \leq c \left(\sum_{T \in \mathcal{T}_h} h_T^{2m} |p|_{H^m(T)}^2\right)^{1/2},$$

where $c > 0$ does not depend on the mesh size h.

4.3.2.5 Basic error estimate

Using the abstract error estimate in Lemma 4.3 and the estimates for the nodal interpolation operators, we can provide a priori bounds for the discretization error.

Theorem 4.25. *Let $\Omega \subset \mathbb{R}^d$ be a polyhedral Lipschitz domain, which is decomposed by a family of shape-regular, quasi-uniform triangulations $(\mathcal{T}_h)$ of mesh size $h \to 0$. Let (σ, u) be the solution to the elasticity problem (3.2) - (3.1) satisfying boundary conditions (3.3) - (3.4). Let (σ_h, u_h) be the Galerkin approximation in $\Sigma_h^k \times V_h^k$ according to Problem 4.11. Assuming that $\sigma \in H_{SYM}^m(\Omega)$ and $u \in [H^{m+1}(\Omega)]^d$ for $1 \leq m \leq k$,*

$$\|\sigma - \sigma_h\|_{\Sigma_h} + \|u - u_h\|_{V_h} \leq c h^m \left(|\sigma|_{H^m(\Omega)} + |u|_{H^{m+1}(\Omega)}\right).$$

Remark 4.26. *The result can easily be generalized to piecewise smooth solutions, then the bound reads*

$$\|\sigma - \sigma_h\|_{\Sigma_h} + \|u - u_h\|_{V_h} \leq c \left(\sum_{T \in \mathcal{T}_h} h_T^{2m} \left(|\sigma|_{H^m(T)}^2 + |u|_{H^{m+1}(T)}^2\right)\right)^{1/2}.$$

When using the approximation spaces $\Sigma_h^{k,k_f} \times V_h^k$ for $k_f := \max(1, k-1)$, the order of convergence is still optimal, as then $P_{SYM}^{k-1}(T) \subset \Sigma_h^{k,k_f}(T)$, and the stability estimate in Lemma 4.17 still holds.

4.3.3 A Korn-type inequality

In Section 4.2.2.6, we decomposed the Nédélec space V_h^k into a global, low-order space $V_{h,0}$, and its high-order complement $\tilde{V}_h^k$. Exact bases for these spaces will be given in Section 4.4; however, we now show a Korn-type inequality on the high-order part $\tilde{V}_h^k$

$$\sum_{T \in \mathcal{T}_h} \|\nabla \tilde{v}_h\|_T^2 \leq c_{K,\tilde{V}}^2 \sum_{T \in \mathcal{T}_h} \|\varepsilon(\tilde{v}_h)\|_T^2 \qquad \forall \tilde{v}_h \in \tilde{V}_h^k.$$

The proof of this inequality is closely related to [Bre04] and additions made in [MW06].

Let now $D \subset \mathbb{R}^d, d = 2, 3$ be a bounded, connected domain satisfying Assumption 3.2. The space of (infinitesimal) rigid body motions $RM(D)$, as defined in (4.31), is exactly the kernel of the strain operator:

$$v \in RM(D) \quad \Leftrightarrow \quad \varepsilon(v) = 0.$$

The low-order space $V_{h,0}$, as defined in Section 4.4, is closely related to the rigid body motions: On each element $T \in \mathcal{T}_h$, the local space $V_{h,0}(T)$ is exactly equal to $RM(T)$. For $v_h \in V_h^k$, we have

$$v_h \in V_{h,0} \quad \Leftrightarrow \quad v_h|_T \in RM(T) \quad \forall T \in \mathcal{T}_h.$$

Thus, $V_{h,0}$ consists of piecewise rigid body motions, which have continuous tangential components.

Let $\mathcal{I}_{h,0}^V$ denote the nodal interpolation operator for the low order Nédélec space $V_{h,0}$, as defined in Section 4.3.2.2. It preserves the piecewise rigid body motions on the tangentially continuous space V_h^k. In Section 4.2.2.6, we decomposed the Nédélec space into a low-order space $V_{h,0}$ and a high order space $\tilde{V}_h^k$. This induces a unique decomposition of $v_h \in V_h^k$

$$v_h = v_0 + \tilde{v} \quad \text{such that} \quad v_0 \in V_{h,0}, \, \tilde{v} \in \tilde{V}_h^k.$$

For this decomposition, we have $\mathcal{I}_{h,0}^V v_h = v_0$.

We will first prove a Korn-type inequality on an element of the triangulation. There, the constant c_T depends on the shape of the element only, and not on its size h. In this proof, we use the fact that the low-order Nédélec interpolant $\mathcal{I}_{h,0}^V$ is stable on V_h^k in the L^2 norm, i.e.

$$\|\mathcal{I}_{h,k}^V v_h\|_\Omega \leq c \|v_h\|_\Omega \qquad \forall v_h \in V_h^k.$$

This estimate holds independently of the mesh size h. One can prove this using the finite overlap and linear independence of the discrete basis functions.

Lemma 4.27. *Let $v_h \in V^k(T)$ for some element $T \in \mathcal{T}_h$. It satisfies the Korn-type inequality*

$$\|\nabla v_h\|_T^2 \leq c_T^2 \left(\|\varepsilon(v_h)\|_T^2 + h_T^{-2} \|\mathcal{I}_{h,0}^V v_h\|_T^2 \right). \tag{4.36}$$

The constant c_T does not depend on the element size h_T.

Proof. We first do the proof on the reference element, where $h_{\hat{T}} \simeq 1$. Assume that the inequality does not hold, i.e. there exists a sequence $(\hat{v}_{h,n})$ in $\hat{V}_k$ such that

$$\|\varepsilon(\hat{v}_{h,n})\|_{\hat{T}}^2 + \|\mathcal{I}_k^{V(\hat{T})}\hat{v}_{h,n}\|_{\hat{T}}^2 \to 0 \quad \text{and} \quad \|\nabla \hat{v}_{h,n}\|_{\hat{T}} = 1 \quad \forall n \in \mathbb{N}.$$

As the interpolation operator $\mathcal{I}_0^{\hat{V}}$ preserves constants, we see by a Poincaré-type inequality that not only $\|\nabla \hat{v}_{h,n}\|_{\hat{T}}$, but also the full norm $\|\hat{v}_{h,n}\|_{H^1(T)}$ is bounded. The compact embedding of H^1 in L^2 (see e.g. [GR86, Wer02]) ensures convergence of a subsequence, without restriction of generality we assume $\hat{v}_{h,n} \to \hat{v}_0$ in L^2. We know that the limit $\hat{v}_0$ lies in $\hat{V}_k$, as the finite-dimensional space is closed in L^2. Using Korn's inequality (Theorem 3.16), we may conclude that $\hat{v}_{h,n} \to \hat{v}_0$ in $H^1(\hat{T})$, and therefore $\|\nabla \hat{v}_0\| = 1$. As $\mathcal{I}_0^{\hat{V}}$ is continuous on $\hat{V}_k$ in the L^2 norm, we obtain

$$\|\varepsilon(\hat{v}_0)\|_{\hat{T}}^2 + \|\mathcal{I}_0^{\hat{V}}\hat{v}_0\|_{\hat{T}}^2 = \lim_{n\to\infty}\left(\|\varepsilon(\hat{v}_{h,n})\|_{\hat{T}}^2 + \|\mathcal{I}_0^{\hat{V}}\hat{v}_{h,n}\|_{\hat{T}}^2\right) = 0.$$

This ensures $\hat{v}_0 = 0$, which contradicts $\|\nabla \hat{v}_0\|_{\hat{T}} = 1$, and thereby concludes the proof on the reference element. The mesh-size dependent estimate (4.36) then follows from a scaling argument, where the conforming transformation Φ_T^V is used. $\square$

We can now prove a Korn-type inequality for $v_h \in V_h^k$, which is based on the splitting of V_h^k into its high-order and low-order components. The proof heavily relies on the techniques developed in [Bre04]. From the proof of [Bre04, Lemma 2.2], one can see that for a piecewise linear vector field $v_0 \in V_{h,0}$ on the quasi-uniform triangulation $\mathcal{T}_h$ there holds

$$\sum_{T \in \mathcal{T}_h} \|\nabla v_0\|_T^2 \leq c \left(\sum_{T \in \mathcal{T}_h} \|\varepsilon(v_0)\|_T^2 + \|v_0\|_\Omega^2 + \sum_{F \in \mathcal{F}_h} h_F^{d-2} \sum_{P \in \mathcal{V}(F)} |[\![v_0]\!]_F(P)|^2 \right). \quad (4.37)$$

There $\mathcal{V}(F)$ denotes the set of vertices of a facet F, and $[\![\cdot]\!]_F$ is the jump operator across this facet (cf. Definition 4.5).

Lemma 4.28. *Let $(\mathcal{T}_h)$ be a family of shape-regular, quasi-uniform triangulations. There exists a constant $\bar{c}_K > 0$ such that for all $v_h \in V_h^k$*

$$\sum_{T \in \mathcal{T}_h} \|\nabla v_h\|_T^2 \leq \bar{c}_K^2 \left(\sum_{T \in \mathcal{T}} \|\varepsilon(v_h)\|_T^2 + \|\mathcal{I}_{h,0}^V v_h\|_\Omega^2 + \sum_{F \in \mathcal{F}_h} h_F^{-1} \|[\![\mathcal{I}_{h,0}^V v_h]\!]_n\|_F^2 \right). \quad (4.38)$$

The constant $\bar{c}_K$ is independent of the local mesh size.

Proof. We use the decomposition

$$\frac{1}{2} \sum_{T \in \mathcal{T}_h} \|\nabla v_h\|_T^2 \leq \sum_{T \in \mathcal{T}_h} \|\nabla(v_h - \mathcal{I}_{h,0}^V v_h)\|_T^2 + \sum_{T \in \mathcal{T}_h} \|\nabla \mathcal{I}_{h,0}^V v_h\|_T^2. \quad (4.39)$$

We estimate both terms on the right hand side separately. For the first term in (4.39), we use

Lemma 4.27, which yields

$$\sum_{T\in\mathcal{T}_h} \|\nabla(v_h - \mathcal{I}_{h,0}^V v_h)\|_T^2 \leq c \sum_{T\in\mathcal{T}_h} \left[\|\varepsilon(v_h - \mathcal{I}_{h,0}^V v_h)\|_T^2 + h_T^{-2}\|\mathcal{I}_{h,0}^V(v_h - \mathcal{I}_{h,0}^V v_h)\|_T^2 \right] = c \sum_{T\in\mathcal{T}_h} \|\varepsilon(v_h)\|_T^2.$$

As $\mathcal{I}_{h,0}^V v_h$ is piecewise linear with $\varepsilon(\mathcal{I}_{h,0}^V v_h) = 0$, the second term in (4.39) can be bounded using inequality (4.37), setting $v_0 = \mathcal{I}_{h,0}^V v_h$:

$$\sum_{T\in\mathcal{T}_h} \|\nabla \mathcal{I}_{h,0}^V v_h\|_T^2 \leq c \left(\|\mathcal{I}_{h,0}^V v_h\|_\Omega^2 + \sum_{F\in\mathcal{F}_h} h_F^{d-2} \sum_{P\in\mathcal{V}(F)} |[\![\mathcal{I}_{h,0}^V v_h]\!]_F(P)|^2 \right).$$

On F, the jump $[\![\mathcal{I}_{h,0}^V v_h]\!]$ across this facet is linear. Moreover, we know that the tangential component of $\mathcal{I}_{h,0}^V v_h$ is continuous. This gives

$$h_F^{d-2} \sum_{P\in\mathcal{V}(F)} |[\![\mathcal{I}_{h,0}^V v_h]\!]_F(P)|^2 \leq c h_F^{-1} \|[\![\mathcal{I}_{h,0}^V v_h]\!]_n\|_F^2.$$

Taking all estimates together, we obtain the required result. □

Corollary 4.29. *Let* $(\mathcal{T}_h)$ *be a family of shape-regular, quasi-uniform triangulations. Let* $\tilde{V}_h^k$ *be the high-order finite element subspace of* V_h^k *according to the splitting in Section 4.2.2.6. Then there exists a constant* $\bar{c}_{K,\tilde{V}}$ *independent of the local mesh size such that for all* $\tilde{v}_h \in \tilde{V}_h^k$

$$\sum_{T\in\mathcal{T}_h} \|\nabla \tilde{v}_h\|_T^2 \leq \bar{c}_{K,\tilde{V}}^2 \sum_{T\in\mathcal{T}_h} \|\varepsilon(\tilde{v}_h)\|_T^2. \tag{4.40}$$

Proof. The result follows directly from the fact that $\mathcal{I}_{h,0}^V \tilde{v}_h = 0$ for $\tilde{v}_h \in \tilde{V}_h^k$. □

4.4 TD-NNS elements

In the sequel, we provide bases for the local finite element spaces $\hat{\Sigma}^k$ and $\hat{V}^k$ on simplicial reference elements in two or three space dimensions, i.e. on triangles and tetrahedrons. Although the suggested bases are not nodal as suggested in Definition 4.8, they will be appropriate for forming the global spaces Σ_h^k, V_h^k. The shape functions are divided into ones corresponding to edges, facets and element interiors. For $\hat{V}^k$, each edge-based shape corresponds to one edge degree of freedom, and vanishes for all others. Also, a facet-bound shape will only take non-zero values for degrees of freedom lying on this facet and the interior of the element, and not for nodes on edges or other facets. Similarly, interior shapes have zero nodal values in all nodes lying on element edges or facets. This way, global finite element shape functions corresponding to an edge, facet or element in the mesh can be built by transforming the respective shape on the reference element to the matching mesh elements.

Orientation In order to obtain a conforming finite element space by transforming the shape functions from the reference element to elements in the mesh, a unique orientation of edges and facets is necessary. In this work, we assume that global vertex numbers of the respective element are available on the reference element. Then the edge between local vertices V_α, V_β is oriented, such that it points from the higher- to the lower-indexed global vertex. We refer to an oriented edge by brackets, $E = [V_\alpha, V_\beta]$, where the first vertex is the one with the higher global index. On an edge E shared between two elements in the mesh, the tangent will point into the same direction on both elements. In three dimensions, we similarly introduce oriented facets, $F = [V_\alpha, V_\beta, V_\gamma]$, where the vertices are again ordered such that their global index is decreasing. A more detailed description of the orientation problem can be found in [Mon03, Zag06]. Another method to come by this problem is described in [AC03], where one defines different bases for the respective orientations of the element in the mesh.

Notation Throughout the following, all calculations are done on the reference element, which we usually denote by $\hat{T}$. However, in order not to complicate notation, we will drop the hat throughout this section, and simply write T. We do similarly for all other involved quantities, especially we use local coordinates x instead of $\hat{x}$, and build finite element spaces Σ^k, V^k instead of $\hat{\Sigma}^k, \hat{V}^k$.

4.4.1 Reference elements and orthogonal polynomials

We define the reference simplices of dimensions one, two and three, i.e. the reference segment T_1, triangle T_2, and tetrahedron T_3. We further provide hierarchical bases for $P^k(T_d)$ as well as $P_0^k(T_d)$ which are orthogonal with respect to the L^2 inner product. We use the families of Legendre and Jacobi polynomials for their construction.

4.4.1.1 The reference segment

The one-dimensional reference element $T_1 = (0, 1)$ is the line segment with vertices $V_1 = 0$ and $V_2 = 1$. We will use the concept of barycentric coordinates, which are linear functions on the segment, and take value one in one vertex, and vanish in the other. We define

$$\lambda_1 := 1 - x_1, \qquad \lambda_2 := x_1.$$

The barycentric coordinates provide a partition of unity, and are used for the standard, lowest order nodal basis ("hat basis") for the discretization of H^1.

On the segment T_1, we provide a basis for $P^k(T_1)$ using *Legendre polynomials*. These polynomials, denoted by $\{\ell_i : 0 \leq i \leq k\}$ form a basis of $P^k[-1, 1]$. They are orthogonal in the L^2 sense, they satisfy

$$\int_{-1}^{1} \ell_i(\xi)\ell_j(\xi)\, d\xi = \frac{2}{2i+1}\delta_{ij}.$$

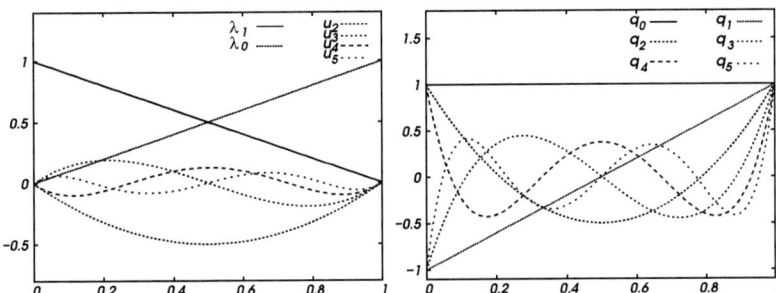

Figure 4.1: The reference segment: left: barycentric coordinates and integrated Legendre polynomials, right: Legendre polynomials.

There are many possibilities for their definition, in computations we use the three-term recurrence relation [Dem07]

$$\ell_0(\xi) = 1,$$
$$\ell_1(\xi) = x,$$
$$(i+1)\ell_{i+1}(\xi) = (2i+1)\ell_i(\xi)\xi - i\ell_{i-1}(\xi) \qquad \text{for } i \geq 2.$$

There exists a similar recurrence for the derivatives of Legendre polynomials. This enables us to evaluate the polynomials as well as their derivatives fast and numerically stable.

A basis for $P^k(T_1)$ is now given by $\{q_i(\lambda_1, \lambda_2) : 0 \leq i \leq k\}$ where

$$q_i(\lambda_\alpha, \lambda_\beta) := \ell_i(\lambda_\beta - \lambda_\alpha). \tag{4.41}$$

This family is orthogonal in the L^2 sense.

The integrated Legendre polynomials $\{L_i : 2 \leq i \leq k\}$ are defined on $[-1, 1]$ by

$$L_i(\xi) := \int_{-1}^{\xi} \ell_{i-1}(\eta) d\eta \qquad \text{for } 2 \leq i \leq k.$$

Due to their definition, these polynomials are orthogonal with respect to the H^1 semi-norm,

$$\int_{-1}^{1} L_i'(\xi) L_j'(\xi) \, d\xi = \frac{2}{2i+1} \delta_{ij}.$$

We will use that the integrated Legendre polynomials vanish at the endpoints $\{-1, 1\}$, and therefore form a basis for $P_0^k[-1, 1]$. This way, the family $\{u_i(\lambda_1, \lambda_2) : 2 \leq i \leq k\}$, where

$$u_i(\lambda_\alpha, \lambda_\beta) := L_i(\lambda_\beta - \lambda_\alpha),$$

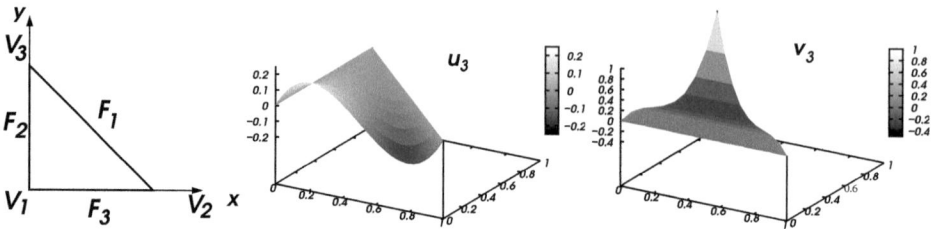

Figure 4.2: The reference triangle: left: sketch of T_2, middle and right: polynomial functions u_3, v_3 on the triangle, generated using the Duffy transform.

is a basis for $P_0^k(T_1)$. For a visualization of the reference element and the respective families of polynomials, see Figure 4.1.

The Legendre polynomials were constructed in such a way, that they are orthogonal with respect to the L^2 inner product. They are a special case of a larger family of polynomials, namely the Jacobi polynomials. We denote *Jacobi polynomials* by

$$p_i^{(\alpha,\beta)}(\xi) \qquad \text{for } \xi \in [-1,1], \; \alpha, \beta > -1.$$

They are orthogonal on $L^2[-1,1]$ with respect to the weight function $w_{\alpha,\beta}(\xi) = (1-\xi)^\alpha (1+\xi)^\beta$. Also Jacobi polynomials can be computed via three-term recurrence relations. We refer to [KS99] for some general properties of Jacobi polynomials, for an extensive overview, including their application in the finite element method, see [Pil08].

4.4.1.2 The reference triangle

In the plane $\mathbb{R}^2$, we define the reference element T_2 to be the triangle with vertices $V_1 = (0,0)$, $V_2 = (1,0)$ and $V_3 = (0,1)$. Its edges E_i, $i = 1, 2, 3$ shall be chosen such that edge E_i lies opposite vertex V_i. As edges and facets coincide in two space dimensions, we set $F_i = E_i$. For a sketch of the reference triangle, see Figure 4.2. We define barycentric coordinates on the triangle to be linear functions, which take value one in one vertex, and vanish in the other two vertices. We set

$$\lambda_1 := 1 - x_1 - x_2, \qquad \lambda_2 := x_1, \qquad \lambda_3 := x_2.$$

Then $\lambda_i(V_i) = 1$ and $\lambda_i|_{E_i} = 0$ for $i = 1, 2, 3$, i.e. the barycentric coordinate matching vertex V_i vanishes on edge E_i. As in the one-dimensional case, they provide a partition of unity and are widely used in finite element methods for H^1 problems.

We need the notion of scaled Legendre and integrated Legendre, as well as scaled Jacobi polynomials. These functions live on the triangle $\{(\xi, \eta) : \eta \in [0,1], \xi \in [-\eta, \eta]\}$, and are generated from tensor-product functions on the unit square using the Duffy transform. According to

[KS99, Zag06], where also a more detailed discussion can be found, we define the following scaled polynomial functions for $i \geq 0, j \geq 2$:

$$\ell_i^S(\xi,\eta) := \eta^i \ell_i\left(\frac{\xi}{\eta}\right), \qquad p_i^{(\alpha,\beta),S}(\xi,\eta) := \eta^i p_i^{(\alpha,\beta)}\left(\frac{\xi}{\eta}\right), \qquad L_j^S(\xi,\eta) := \eta^j L_j\left(\frac{\xi}{\eta}\right).$$

Basically, polynomial functions on the edge are extended to the interior of the triangle. We use the family $\{q_{ij}(\lambda_1, \lambda_2, \lambda_3) : 0 \leq i+j \leq k\}$, where

$$q_{ij}(\lambda_\alpha, \lambda_\beta, \lambda_\gamma) := \ell_i^S(\lambda_\beta - \lambda_\alpha, \lambda_\alpha + \lambda_\beta) p_j^{(2i+1,0)}(\lambda_\gamma - \lambda_\alpha - \lambda_\beta). \tag{4.42}$$

In this definition, and also in the following, we inherently assume that indices only take values in the range of definition of the respective quantity, i.e. $i, j \geq 0$ in this case. This family spans $P^k(T_2)$ and is linearly independent. Note that the restriction of $q_{ij}(\lambda_\alpha, \lambda_\beta, \lambda_\gamma)$ to edge E_γ coincides with $q_i(\lambda_\alpha, \lambda_\beta)$ from the one-dimensional element. This basis is orthogonal with respect to the L^2 inner product, as one can easily see from the defining properties of Legendre and Jacobi polynomials.

For the polynomial space satisfying zero boundary conditions, let us define

$$u_i(\lambda_\alpha, \lambda_\beta) := L_i^s(\lambda_\beta - \lambda_\alpha, \lambda_\alpha + \lambda_\beta) \qquad i \geq 2,$$
$$v_j(\lambda_\alpha, \lambda_\beta, \lambda_\gamma) := \lambda_\gamma \ell_{j-1}(\lambda_\gamma - \lambda_\alpha - \lambda_\beta) \qquad j \geq 1.$$

There, u_i vanishes on edges E_α, E_β, and is a polynomial of order i along edge $E_\gamma = [V_\alpha, V_\beta]$. Indeed, it is defined in such a way such that $u_i(\lambda_\alpha, \lambda_\beta)$ restricted to edge E_γ coincides with the bubble function u_i defined on the reference segment. The second family of polynomials, namely $\{v_j\}$, takes zero values on the third edge E_γ. We plot two functions u_i, v_j in Figure 4.2. We can characterize $P_0^k(T_2)$ as

$$P_0^k(T_2) = \operatorname{span}\{u_i(\lambda_1, \lambda_2) v_j(\lambda_1, \lambda_2, \lambda_3) : 3 \leq i+j \leq k\}.$$

4.4.1.3 The reference tetrahedron

Let now the reference tetrahedron T_3 be defined as the convex hull of its vertices, $T_3 = [V_1, V_2, V_3, V_4]$, where $V_1 = (0,0,0)$, $V_2 = (1,0,0)$, $V_3 = (0,1,0)$ and $V_4 = (0,0,1)$, see also Figure 4.3. Then, the barycentric coordinates are given by

$$\lambda_1 := 1 - x_1 - x_2 - x_3, \qquad \lambda_2 := x_1, \qquad \lambda_3 := x_2, \qquad \lambda_4 := x_3.$$

Similar to the one- and two-dimensional elements, we first provide a basis for the full polynomial space of order k on the tetrahedron. This is realized by the family $\{q_{ijl} : 0 \leq i+j+l \leq k\}$, where

$$q_{ijl}(\lambda_\alpha, \lambda_\beta, \lambda_\gamma, \lambda_\delta) :=$$
$$\ell_i^S(\lambda_\beta - \lambda_\alpha, \lambda_\alpha + \lambda_\beta) \, p_j^{(2i+1,0),S}(\lambda_\gamma - \lambda_\alpha - \lambda_\beta, \lambda_\alpha + \lambda_\beta + \lambda_\gamma) \, p_l^{(2i+2j+2)}(\lambda_\delta - \lambda_\alpha - \lambda_\beta - \lambda_\gamma).$$

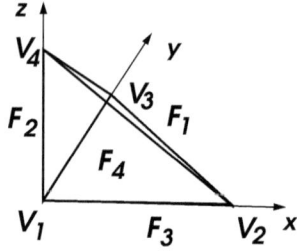

Figure 4.3: Reference tetrahedron.

This family is constructed such that, restricting q_{ijl} to edge $E = [V_\alpha, V_\beta]$, one obtains $q_i(\lambda_\alpha, \lambda_\beta)$ on the segment. On the face $F_\delta = [V_\alpha, V_\beta, V_\gamma]$, q_{ijl} coincides with q_{ij} as defined on the triangle. Again, the family is orthogonal with respect to the L^2 inner product.

Last, we come to the polynomial space of order k satisfying homogenous boundary conditions. We set

$$u_i(\lambda_\alpha, \lambda_\beta) := L_i^s(\lambda_\beta - \lambda_\alpha, \lambda_\alpha + \lambda_\beta) \qquad i \geq 2,$$
$$v_j(\lambda_\alpha, \lambda_\beta, \lambda_\gamma) := \lambda_\gamma \ell_{j-1}^S(\lambda_\gamma - \lambda_\alpha - \lambda_\beta, \lambda_\alpha + \lambda_\beta + \lambda_\gamma) \qquad j \geq 1,$$
$$w_l(\lambda_\alpha, \lambda_\beta, \lambda_\gamma, \lambda_\delta) := \lambda_\delta \ell_{l-1}(\lambda_\delta - \lambda_\alpha - \lambda_\beta - \lambda_\gamma) \qquad l \geq 1.$$

Here, u_i vanishes on facets F_α, F_β; its restriction to the edge $E = [V_\alpha, V_\beta]$ or to facets containing this edge yields u_i defined on the segment or the triangle, respectively. The family $\{v_j\}$ is zero on facet F_γ due to the factor λ_γ. On the other three facets, it equals v_j defined on the triangle. Finally, w_l is defined such that it vanishes on the fourth facet F_δ. Together, they can be used as a basis for $P_0^k(T_3)$:

$$P_0^k(T_3) = \text{span}\{u_i v_j w_l : 4 \leq i + j + l \leq k\}.$$

4.4.2 On the triangle

We now define shape functions for V^k and Σ^k on the triangle $T := T_2$.

4.4.2.1 A triangular element for the Nédélec space

Hierarchical constructions of finite elements for the Nédélec space are well known. We use the shape functions provided in [Zag06], where also a detailed analysis of their properties is given.

As already mentioned, the finite element shape functions are organized in edge- and cell-based functions. This is realized in the following way: Let $V^k := \text{span } \Psi_k^V$, where

$$\Psi_k^V := \left(\bigcup_{E \in \mathcal{E}(T)} \Psi_{E,k}^V \right) \cup \Psi_{T,k}^V.$$

Then $\Psi_{E,k}^V$ shall consist of shapes bound to edge E, whereas $\Psi_{T,k}^V$ corresponds to the interior shapes. We define

- the set of edge-based shapes $\Psi_{E,k}^V := \{\psi_{E,i}^V\}$ for an edge $E = [V_\alpha, V_\beta]$

$$\psi_{E,0}^V := \nabla \lambda_\alpha \lambda_\beta - \lambda_\alpha \nabla \lambda_\beta,$$
$$\psi_{E,i}^V := \nabla u_{i+1}(\lambda_\alpha, \lambda_\beta) \qquad 1 \leq i \leq k,$$

where $\psi_{E,0}^V$ is the lowest order shape function as in [Néd80],

- the set of cell-based shapes $\Psi_{T,k}^V := \{\psi_{T,m,ij}^V\}$

$$\psi_{T,1,ij}^V := \nabla(u_i v_j), \qquad 3 \leq i+j \leq k+1,$$
$$\psi_{T,2,ij}^V := \nabla u_i v_j - u_i \nabla v_j, \qquad 3 \leq i+j \leq k+1,$$
$$\psi_{T,3,0j}^V := (\nabla \lambda_1 \lambda_2 - \lambda_1 \nabla \lambda_2) v_j, \qquad 1 \leq j \leq k-1.$$

Note that the tangential components of $\Psi_{E,k}^V$ span $P^k(E)$ on this edge, whereas they vanish on all other edges. For the interior shapes, the tangential component equals zero on the whole boundary. Therefore, $\Psi_{T,k}^V$ is a basis of $P_{0,\tau}^k(T)$.

In Section 4.2.2.6, we described a decomposition of the Nédélec space into a global, low-order space $V_{h,0}$, and local high-order spaces $\tilde{V}_X^k$ for any X in the union $\mathcal{X}_h = \mathcal{E}_h \cup \mathcal{F}_h \cup \mathcal{T}_h$. Note that, in two space dimensions, $\mathcal{E}_h$ and $\mathcal{F}_h$ coincide. Using the basis functions from above, we can now explicitely define

$$V_{h,0} := \text{span}\left(\bigcup_{E \in \mathcal{E}_h} \{\psi_{E,0}^V\} \right), \qquad (4.43)$$
$$\tilde{V}_E^k := \text{span}\{\psi_{E,i}^V : i = 1, \ldots, k\}, \qquad (4.44)$$
$$\tilde{V}_T^k := \text{span } \Psi_{T,k}^V. \qquad (4.45)$$

4.4.2.2 A triangular element for the stress space

For the definition of finite element shapes for the normal-normal continuous space Σ^k, we first present a set of constant, symmetric tensor fields $\{S^F : F \in \mathcal{F}(T)\}$, such that

- the normal-normal component of S^F vanishes on all facets but F,
- the normal-normal component of S^F on F is constant, and

- the union $\{S^F : F \in \mathcal{F}(T)\}$ spans the space of piecewise constant, symmetric tensor fields $P^0_{SYM}(T)$.

We propose
$$S^{F_i} := \nabla \lambda^\perp_{i+1} \otimes \nabla \lambda^\perp_{i+2}.$$
There the index is to be understood modulo 3. Calculating the fields, one obtains
$$S^{F_1} = \begin{pmatrix} 0 & 1 \\ 1 & 0 \end{pmatrix}, \quad S^{F_2} = \begin{pmatrix} -2 & 1 \\ 1 & 0 \end{pmatrix}, \quad S^{F_3} = \begin{pmatrix} 0 & -1 \\ -1 & 2 \end{pmatrix}. \tag{4.46}$$

Using these tensor fields, it is easy to construct "bubble functions" B^i, for which the normal-normal component vanishes on the whole boundary ∂T. This is simply done by multiplication of S^{F_i} with λ_i, which is zero on facet F_i:
$$B^i := \lambda_i S^{F_i}.$$

Now, we provide a finite element basis Ψ^Σ_k for Σ^k. We organize it in facet- and cell-based functions,
$$\Psi^\Sigma_k := \left(\bigcup_{F \in \mathcal{F}(T)} \Psi^\Sigma_{F,k} \right) \cup \Psi^\Sigma_{T,k}.$$

There we define

- the set of facet-based shapes $\Psi^\Sigma_{F,k} := \{\psi^\Sigma_{F,i}\}$ for a facet $F = [V_\alpha, V_\beta]$
$$\psi^\Sigma_{F,i} := q_{i0}(\lambda_\alpha, \lambda_\beta) S^F, \qquad 0 \leq i \leq k.$$

- the set of cell-based shapes $\Psi^\Sigma_{T,k} := \{\psi^\Sigma_{T,m,ij}\}$
$$\psi^\Sigma_{T,m,ij} := q_{ij} B^m = q_{ij} \lambda_m S^{F_m}, \qquad 0 \leq i+j \leq k-1, \ 1 \leq m \leq 3.$$

The construction ensures that a shape from $\Psi^\Sigma_{F,k}$ takes values only on facet F and not on the other two facets. On F, $\Psi^\Sigma_{F,k}$ spans the full polynomial space $P^k(F)$. Any shape in $\Psi^\Sigma_{T,k}$, which is bound to the interior, satisfies homogenous normal-normal boundary conditions on ∂T. We have $P^k_{0,nn}(T) = \text{span}\left(\Psi^\Sigma_{T,k} \right)$, as the next lemma states. There we show that Ψ^Σ_k really provides a basis for the finite element space Σ^k, i.e. the shapes are linearly independent.

Lemma 4.30. *For an integer $k \geq 0$, the set of shape functions Ψ^Σ_k on the triangle T is linearly independent, and forms a basis for $\Sigma^k = P^k_{SYM}(T)$. The subset $\Psi^\Sigma_{T,k}$ spans the bubble space $P^k_{0,nn}(T)$.*

Proof. We first show the linear independence of Ψ^Σ_k. From the definitions of $\Psi^\Sigma_{F,k}$, $\Psi^\Sigma_{T,k}$ we see that
$$\Psi^\Sigma_k = \bigcup_{m=1}^{3} \{q_{ij} \lambda_m S^{F_m} : 0 \leq i+j \leq k-1\} \cup \bigcup_{m=1}^{3} \{q_{i0} S^{F_m} : 0 \leq i \leq k\}.$$

As, for $m = 1, 2, 3$,

$$\operatorname{span}\{\lambda_m q_{ij} : 0 \leq i+j \leq k-1\} =$$
$$= \operatorname{span}\{\lambda_m \ell_i^S(\lambda_{m+2} - \lambda_{m+1}, \lambda_{m+1} + \lambda_{m+2})\ell_j(\lambda_m - \lambda_{m+1} - \lambda_{m+2}) : 0 \leq i+j \leq k-1\}$$
$$= \operatorname{span}\{\ell_i^S(\lambda_{m+2} - \lambda_{m+1}, \lambda_{m+1} + \lambda_{m+2})\ell_{j+1}(\lambda_m - \lambda_{m+1} - \lambda_{m+2}) : 0 \leq i+j \leq k-1\}$$
$$= \operatorname{span}\{q_{i,j+1} : 0 \leq i+j \leq k-1\},$$

we conclude

$$\operatorname{span}\left(\Psi_k^\Sigma\right) = \operatorname{span}\left(\bigcup_{m=1}^3 \{q_{ij} S^{F_m} : 0 \leq i+j \leq k\}\right) = P_{SYM}^k(T).$$

Here we used that the S^F span $P_{SYM}^0(T)$. Counting the respective numbers of shape functions, we get that $|\Psi_k^\Sigma| = 3(k+1)(k+2)/2$, which is exactly the dimension of $P_{SYM}^k(T)$. Analogously, we see that $|\Psi_{T,k}^\Sigma| = 3(k+1)k/2$, which is the dimension of $P_{0,nn}^k(T)$. As the shapes all lie in $P_{SYM}^k(T)$, $P_{0,nn}^k(T)$ respectively, this completes the proof. □

4.4.3 On the tetrahedron

In the following, we define shape functions for V^k, and Σ^k on the tetrahedron $T := T_3$. All constructions are similar to the case of two space dimensions, but of course more involved. Again, we use the basis for the Nédélec space, which was provided and analyzed in [Zag06]. For the stress space, we construct shapes similar to the ones on the triangle.

4.4.3.1 A tetrahedral element for the Nédélec space

On the tetrahedron, shape functions for V^k are divided into edge-, facet- and cell-based ones. We denote the local basis of V^k by Ψ_k^V, which is then decomposed as

$$\Psi_k^V := \left(\bigcup_{E \in \mathcal{E}(T)} \Psi_{E,k}^V\right) \cup \left(\bigcup_{F \in \mathcal{F}(T)} \Psi_{F,k}^V\right) \cup \Psi_{T,k}^V.$$

The set $\Psi_{E,k}^V$ shall consist of shapes bound to edge E, whereas $\Psi_{F,k}^V$ corresponds to a facet F, and $\Psi_{T,k}^V$ contains the interior shapes. We define

- the set of edge-based shapes $\Psi_{E,k}^V := \{\psi_{E,i}^V\}$ for an edge $E = [V_\alpha, V_\beta]$

$$\psi_{E,0}^V := \nabla \lambda_\alpha \lambda_\beta - \lambda_\alpha \nabla \lambda_\beta,$$
$$\psi_{E,i}^V := \nabla u_{i+1}(\lambda_\alpha, \lambda_\beta), \qquad 1 \leq i \leq k,$$

where $\psi_{E,0}^V$ is the lowest order shape function as in [Néd80],

- the set of facet-based shapes $\Psi_{F,k}^V := \{\psi_{F,m,ij}^V\}$ for a facet $F = [V_\alpha, V_\beta, V_\gamma]$

$$\begin{aligned}
\psi_{F,1,ij}^V &:= \nabla(u_i v_j), & 3 \leq i+j \leq k+1, \\
\psi_{F,2,ij}^V &:= \nabla u_i v_j - u_i \nabla v_j, & 3 \leq i+j \leq k+1, \\
\psi_{F,3,0j}^V &:= (\nabla \lambda_1 \lambda_2 - \lambda_1 \nabla \lambda_2) v_j, & 1 \leq j \leq k-1,
\end{aligned}$$

where $u_i = u_i(\lambda_\alpha, \lambda_\beta)$ and $v_j = v_j(\lambda_\alpha, \lambda_\beta, \lambda_\gamma)$,

- the set of cell-based shapes $\Psi_{T,k}^V := \{\psi_{T,m,ijl}^V\}$

$$\begin{aligned}
\psi_{T,1,ijl}^V &:= \nabla(u_i v_j w_l), & 4 \leq i+j+l \leq k+1, \\
\psi_{T,2,ijl}^V &:= \nabla u_i v_j w_l - u_i \nabla v_j w_l + u_i v_j \nabla w_l, & 4 \leq i+j+l \leq k+1, \\
\psi_{T,3,ijl}^V &:= \nabla u_i v_j w_l + u_i \nabla v_j w_l - u_i v_j \nabla w_l, & 4 \leq i+j+l \leq k+1, \\
\psi_{T,4,0jl}^V &:= (\nabla \lambda_1 \lambda_2 - \lambda_1 \nabla \lambda_2) v_j w_l, & 2 \leq j+l \leq k-1,
\end{aligned}$$

where $u_i = u_i(\lambda_1, \lambda_2)$, $v_j = v_j(\lambda_1, \lambda_2, \lambda_3)$ and $w_l = w_l(\lambda_1, \lambda_2, \lambda_3, \lambda_4)$.

As on the triangle, the tangential components of $\Psi_{E,k}^V$ span $P^k(E)$ on this edge, whereas they vanish on all other edges. Also, shapes bound to facet F have zero tangential trace on all other facets, and for interior shapes, the tangential component is zero on the whole boundary.

Again, we can now define the decomposition of the Nédélec space into a global, low-order space $V_{h,0}$, and local high-order spaces $\tilde{V}_X^k$ for any X in the union $\mathcal{X}_h = \mathcal{E}_h \cup \mathcal{F}_h \cup \mathcal{T}_h$, as was described in Section 4.2.2.6. Using the basis functions from above, we set, similarly to the two-dimensional case,

$$V_{h,0} := \mathrm{span}\left(\bigcup_{E \in \mathcal{E}_h} \{\psi_{E,0}^V\}\right), \tag{4.47}$$

$$\tilde{V}_E^k := \mathrm{span}\{\psi_{E,i}^V : i = 1, \ldots, k\}, \tag{4.48}$$

$$\tilde{V}_F^k := \mathrm{span}\, \Psi_{F,k}^V, \tag{4.49}$$

$$\tilde{V}_T^k := \mathrm{span}\, \Psi_{T,k}^V. \tag{4.50}$$

4.4.3.2 A tetrahedral element for the stress space

Now, to define a basis for Σ^k, we again use constant, symmetric tensor fields, for which the normal-normal component vanishes on all facets but one. The fields

$$S^{F_i} = \mathrm{sym}[(\nabla \lambda_{i+1} \times \nabla \lambda_{i+2}) \otimes (\nabla \lambda_{i+2} \times \nabla \lambda_{i+3})], \quad i = 1, \ldots, 4.$$

satisfy this assumption. In two dimensions, we have seen that the fields S^{F_i} span the space of constant symmetric tensor fields. On the tetrahedron, we have to add two further fields $S^{T,1}$ and $S^{T,2}$, as the corresponding space is of dimension 6. We construct them in such a way, that their normal-normal components vanish on ∂T,

$$S^{T,1} = \text{sym}[(\nabla\lambda_1 \times \nabla\lambda_2) \otimes (\nabla\lambda_3 \times \nabla\lambda_4)],$$
$$S^{T,2} = \text{sym}[(\nabla\lambda_1 \times \nabla\lambda_3) \otimes (\nabla\lambda_2 \times \nabla\lambda_4)].$$

Explicitely, they read as

$$S^{F_1} = \begin{pmatrix} -2 & 1 & 0 \\ 1 & 0 & 0 \\ 0 & 0 & 0 \end{pmatrix}, \quad S^{F_2} = \begin{pmatrix} 0 & 1 & -1 \\ 1 & -2 & 1 \\ -1 & 1 & 0 \end{pmatrix}, \quad S^{F_3} = \begin{pmatrix} 0 & 0 & 0 \\ 0 & 0 & -1 \\ 0 & -1 & 2 \end{pmatrix},$$

$$S^{F_4} = \begin{pmatrix} 0 & 0 & 1 \\ 0 & 0 & 0 \\ 1 & 0 & 0 \end{pmatrix}, \quad S^{T,1} = \begin{pmatrix} 0 & 0 & -1 \\ 0 & 0 & 1 \\ -1 & 1 & 0 \end{pmatrix}, \quad S^{T,2} = \begin{pmatrix} 0 & -1 & 0 \\ -1 & 0 & 1 \\ 0 & 1 & 0 \end{pmatrix}.$$

As for the triangle, we additionally provide bubble functions B^i. For the four facets F^i, we set

$$B^i := \lambda_i S^{F_i}.$$

Additionally, we use

$$B^5 = S^{T,1}, \qquad B^6 = S^{T,2},$$

as these fields are already element bubbles.

We define a finite element basis Ψ_k^Σ for Σ^k. As on the triangle, we split it into facet- and cell-based functions,

$$\Psi_k^\Sigma := \left(\bigcup_{F \in \mathcal{F}(T)} \Psi_{F,k}^\Sigma \right) \cup \Psi_{T,k}^\Sigma.$$

There we define

- the set of facet-based shapes $\Psi_{F,k}^\Sigma := \{\psi_{F,ij}^\Sigma\}$ for a facet $F = [V_\alpha, V_\beta, V_\gamma]$

$$\psi_{F,i}^\Sigma := q_{ij0}(\lambda_\alpha, \lambda_\beta, \lambda_\gamma) S^F, \qquad 0 \leq i+j \leq k,$$

- the set of cell-based shapes $\Psi_{T,k}^\Sigma := \{\psi_{T,m,ijk}^\Sigma\}$

$$\psi_{T,m,ijk}^\Sigma := q_{ij} B^m = q_{ij} \lambda_m S^{F_m}, \qquad 0 \leq i+j+l \leq k-1,\ 1 \leq m \leq 4,$$
$$\psi_{T,m,ijk}^\Sigma := q_{ij} B^m = q_{ij} S^{T,m-4}, \qquad 0 \leq i+j+l \leq k,\ m = 5,6.$$

From this construction, one can see that the facet-based shapes satisfy zero normal-normal boundary conditions on all facets but one, whereas the normal-normal components of interior shapes are totally vanishing. Similar to the triangle, one can prove the following lemma, which states that the spaces of shape functions are well chosen.

Lemma 4.31. *For an integer $k \geq 0$, the set of shape functions Ψ_k^Σ on the tetrahedron T is linearly independent, and forms a basis for $\Sigma^k = P_{SYM}^k(T)$. The subset $\Psi_{T,k}^\Sigma$ spans the bubble space $P_{0,nn}^k(T)$.*

4.5 Numerical results

In the following, we present results obtained from computations with the TD-NNS method. We use the open-source software package Netgen/NgSolve, see http://www.hpfem.jku.at or [Sch97]. For the examples described below, we implemented stress elements of arbitrary order on triangles and tetrahedrons, as are described in Section 4.4.

The first example concerns computations on the unit square. We assume that the left hand side is fixed (i.e. it corresponds to Γ_D, where $u = u_D = 0$). Surface forces are prescribed on the right hand side, where we set $\sigma_n = (1,1)^T$. The remaining parts of the boundary are free, i.e. they are governed by homogenous boundary conditions. We assume to have a homogenous, isotropic, linearly elastic material with Young's modulus $\bar E = 1$, and Poisson ratio $\bar\nu = 0.4$. We use an initial mesh consisting of 18 triangular elements, and calculate with elements of uniform orders $k = 1, 2, 4$. We do adaptive refinement as implied by a Zienkiewicz-Zhu type error estimator, which was introduced for H^1 problems in [ZZ87]. In Figure 4.4, we plot the convergence of the stress in the L^2 norm. We observe the predicted optimal order of convergence. Additionally, we consider an hp-version of the finite element method. We prescribe an a-priori geometric mesh refinement towards the singularities occurring in the corners of the domain, and then increase the order of the finite element method. In Figure 4.4 we plot the results obtained for five levels of geometric refinement.

As a second example, we consider a crankshaft in three space dimensions. We assume that the shaft is fixed on one end, whereas a tangential force is acting counter-clock-wise on the other end. The remaining boundary is free. We assume the shaft is made of steel, with a Poisson ratio of $\bar\nu = 0.3$ and a Young's modulus $\bar E = 2 \cdot 10^{11} N/m^2$. We use a mesh consisting of 6 710 tetrahedral elements, which are partially curved to approximate the geometry of the shaft. We use a method of order $k = 4$, with facet-order $k_f = 3$. The corresponding linear system contains 440 725 coupling degrees of freedom. In Figure 4.5, we plot the stress components $\sigma_{x_2 x_3}$ and $\sigma_{x_2 x_2}$.

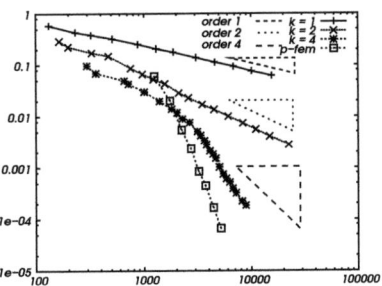

Figure 4.4: Coupling degrees of freedom vs. error $\|\sigma - \sigma_h\|_\Omega$: We plot the convergence of methods of orders $k = 1, 2, 4$, as well as of the p-version on an a-priori geometrically refined grid.

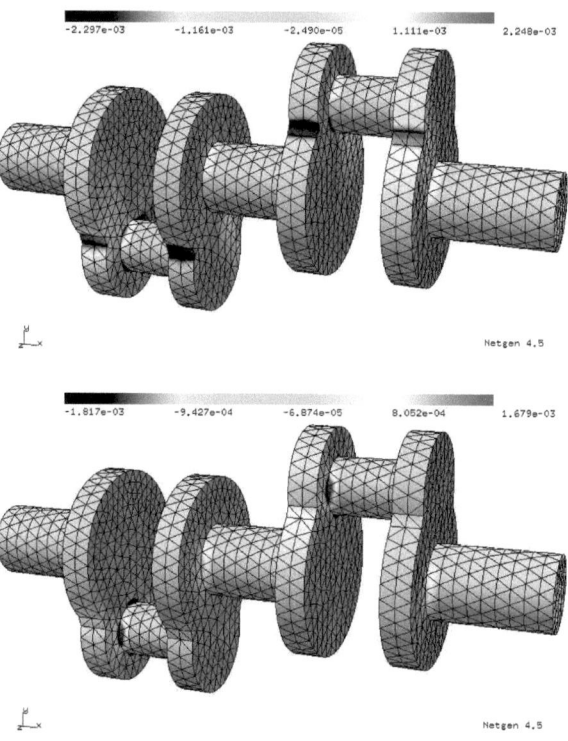

Figure 4.5: Crankshaft, 6710 elements, order $k = 4$, 440 725 coupling degrees of freedom. We plot the stress components $\sigma_{x_2 x_3}$ (upper plot) and $\sigma_{x_2 x_2}$ (lower plot).

Chapter 5
A hybrid finite element method

The present chapter is devoted to hybridization techniques for the TD-NNS method. In general, hybridization of a mixed method means that the imposed inter-element continuity of the flux field (which is the stress field in context of elasticity) is not enforced directly by a conforming choice of the finite element space, but via Lagrangian multipliers living on element interfaces. In our case, we see that these multipliers correspond to normal displacements on the facets. This hybridized problem is equivalent to the original one in the sense that the calculated stresses and displacements coincide. We can directly transfer stability properties of the TD-NNS method to this formulation.

The Lagrangian multipliers enter the set of equations as additional unknowns. At first glance, it does not seem feasible to enlarge the number of unknowns deliberately. Subsequently, we present several reasons in favor of hybridization. One of them is the simpler structure of the stress finite element space. Instead of the normal-normal continuous finite element basis described in Section 4.4, one can use piecewise polynomial, globally discontinuous shape functions.

A major benefit of the hybrid setup is that all degrees of freedom for the stresses are completely local to each element. Thus, the stress field can be eliminated from the system of equations at optimal complexity. One is left with solving a system for the displacement quantities, i.e. for u and the Lagrangian multiplier only. The remaining system matrix is symmetric and positive definite (SPD). Therefore, iterative solvers such as the conjugate gradient (CG) method are directly applicable. We propose an additive Schwarz block preconditioner, which is spectrally equivalent to the inverse of the system matrix, independently of the mesh size h. With this choice, the preconditioned conjugate gradient (PCG) method is an optimal solver for the elasticity problem.

Moreover, we are concerned with finding a finite element method, which is stable also in the case of almost incompressible materials. Assuming that the material is linear elastic, isotropic and homogenous, we call it almost incompressible, if the Poisson ratio $\bar\nu$ approaches $\frac{1}{2}$, or, equivalently, the Lamé parameter $\bar\lambda$ tends to infinity. We have seen that in this case, all stability estimates for the primal variational formulation deteriorate, whereas stability can be preserved for the mixed scheme of Hellinger and Reissner. Also our new mixed method is stable in the infinite-dimensional setting. However, we have lost this property when doing a straightforward discretization as in Chapter 4.

We analyze the problem in the hybrid setup using discrete, parameter dependent norms. We prove stability of the hybridized formulation with respect to these norms, and error estimates for the displacement, which are independent of the degree of incompressibility. Moreover, we can show that our additive Schwarz block preconditioner is spectrally equivalent to the system matrix also for nearly incompressible materials, and thus the corresponding PCG algorithm is an optimal choice for solving the discrete system.

In Section 5.1, we are concerned with hybridization methods in general and their application to the TD-NNS method. We verify stability estimates for the hybridized method and address implementation issues. Section 5.2 deals with the construction of a block preconditioner, which is well suited in a PCG method applied to the hybrid Schur complement problem. Finally, we apply these techniques to nearly incompressible materials in Section 5.3, which are confirmed by the examples in Section 5.4.

5.1 Hybridization of the TD-NNS method

In this section, we give a short introduction into hybridization methods, and apply this technique to the TD-NNS problem. We provide stability and error estimates for the hybridized system.

5.1.1 Hybridization basics

The main idea of hybridization is replacing a globally continuous finite element space by local spaces, which are associated to a prescribed, non-overlapping domain decomposition. As the continuity across interfaces of the domain decomposition is broken in this process, it is re-enforced by Lagrangian multipliers. These methods are closely related to discontinuous Galerkin (DG) methods. In context of the TD-NNS method, we cut the inter-elemental, normal-normal continuity of the stresses, and re-enforce it by Lagrangian multipliers. These multipliers resemble normal displacements on element facets.

Originally, hybridization methods were introduced as a technique of implementing mixed finite elements, first done in [dV65]. In this setting, the domain decomposition mentioned above is simply the finite element triangulation $\mathcal{T}_h$ of Ω. In [AB85], hybridization of Raviart-Thomas elements in the mixed method for the Poisson equation, as well as hybridization for a fourth order problem are discussed. There, it is shown that the solution of the mixed and the hybridized problem coincide, and an a-priori error bound for the Lagrangian parameter is given. We refer to [CG04], where hybridization of Raviart-Thomas and Brezzi-Douglas-Marini finite element methods are analyzed, and to [CG05c], where the authors provide error analysis for the Lagrangian multiplier in variable degree spaces. In [BF91], a more extensive introduction to hybrid methods can be found.

Hybridization has also been used for the discretization of the Stokes problem. There, one uses these techniques to find finite element functions which are incompressible, i.e. divergence free. We mention [CG05b, CG05a, CCS06], where this goal is achieved in both two and three space dimensions.

A major advantage of hybridization is, that we regain a symmetric, positive definite system matrix, as is characteristic for the primal problem. This property is usually lost when mixed formulations are considered. In the following, we want to sketch this issue in a little more detail. The primal problem (the pure displacement formulation in our case) corresponds to a problem of energy minimization, whereas mixed formulations (e.g. the primal mixed method, the Hellinger-Reissner formulation, or the TD-NNS method) are equivalent to saddle point problems. The corresponding system matrix for the discretized problem is thus indefinite; in general, one ends up with an operator equation of the form

$$\begin{pmatrix} A_h & B_h^* \\ B_h & 0 \end{pmatrix} \begin{pmatrix} \sigma_h \\ u_h \end{pmatrix} = \begin{pmatrix} F_1 \\ F_2 \end{pmatrix}, \tag{5.1}$$

where the asterisk denotes the adjoint operator. In this formulation, the required continuity properties of the flux variable σ_h is enforced directly by a conforming choice of the solution space Σ_h. In hybridization methods, this continuity is broken, and re-enforced via a Lagrangian multiplier λ_h, which lives on the facets of the triangulation. The structure of equation (5.1) changes to

$$\begin{pmatrix} A_h & B_{1,h}^* & B_{2,h}^* \\ B_{1,h} & 0 & 0 \\ B_{2,h} & 0 & 0 \end{pmatrix} \begin{pmatrix} \sigma_h \\ u_h \\ \lambda_h \end{pmatrix} = \begin{pmatrix} F_1 \\ F_2 \\ 0 \end{pmatrix}. \tag{5.2}$$

Due to the fact that the flux σ_h is left discontinuous, the upper left block A_h is block-diagonal, where each block corresponds to the element matrix of one single element. Therefore, A_h can be inverted in optimal complexity, and thus σ_h is eliminated from system (5.2). We arrive at the Schur complement system

$$S_h \begin{pmatrix} u_h \\ \lambda_h \end{pmatrix} = H_h, \tag{5.3}$$

where

$$S_h := \begin{pmatrix} B_{h,1} \\ B_{h,2} \end{pmatrix} A_h^{-1} (B_{h,1}^*, B_{h,2}^*), \quad \text{and} \quad H_h := \begin{pmatrix} B_{h,1} \\ B_{h,2} \end{pmatrix} A_h^{-1} F_1 - \begin{pmatrix} F_2 \\ 0 \end{pmatrix}.$$

This remaining system is then symmetric positive definite, which is feasible when using iterative solvers, such as preconditioned conjugate gradient (PCG) methods. The additional information supplied by the Lagrangian multipliers can even be used for local post-processing, thereby enhancing the solution [AB85, BDM85].

Hybridization methods are closely linked to discontinuous Galerkin (DG) methods. There, one usually starts not from a mixed formulation, but from the primal one, as e.g. the pure displacement formulation. Then, one chooses a piecewise defined, infinite-dimensional but discontinuous space and a corresponding variational formulation, and applies a Galerkin method with respect to some finite-dimensional subspaces (hence the name DG method). The coupling between elements can then occur in various ways, and is usually imposed via *jump* and *stabilization terms*. An overview and analysis on stabilization mechanisms in DG can be found in [BCMS06] For a detailed overview and comprehensive analysis on discontinuous Galerkin methods, we refer to [ABCM02]. DG methods are very flexible when it comes to the choice of approximation spaces, as they can be defined totally local to each element. However, one usually ends up with a high number of globally coupling degrees of freedom. In [CGL07], it was shown that a certain class of DG methods can be hybridized, i.e. be embedded in the framework presented above. This leads to much sparser system matrices with fewer globally coupling dofs.

5.1.2 The hybrid TD-NNS method

So far, we used conforming approximation spaces Σ_h^k, V_h^k, as defined in (4.13) and (4.14), satisfying certain continuity conditions. For the stress space, normal-normal continuity of $\tau_h \in \Sigma_h^k$ was enforced directly. In the spirit of hybridization, we break this continuity, which results in the discontinuous space $\widetilde{\Sigma}_h^k$

$$\widetilde{\Sigma}_h^k := \{ \tau_h \in L^2_{SYM}(\Omega) : \tau_h \in P^k_{SYM}(\mathcal{T}_h) \}. \tag{5.4}$$

Implementing this space in a computational method is much easier than for the original space Σ_h^k, where one has to take care of normal-normal continuity. Now, for the Lagrangian multipliers, we introduce the *facet space*

$$\Lambda_h^k := \{ \mu_h \in L^2(\mathcal{F}_h) : \mu_h \in P^k(\mathcal{F}_h),\ \mu_h|_{\Gamma_D} = 0 \}. \tag{5.5}$$

The realization of this facet space in practice is straightforward: on each facet F, one needs to provide a basis for $P^k(F)$, which can be done using the families of Legendre/Jacobi polynomials described in Section 4.4.1. We comment on these issues in more detail in Section 5.1.4.

Notation We already mentioned that $\lambda_h \in \Lambda_h^k$ plays the role of a flux approximation to the normal displacement u_n. Let $F \in \mathcal{F}_h$ be an arbitrary interior facet, with neighboring elements T_1, T_2. Without restriction of generality, we assume that the normal n_F of the oriented facet F coincides with the outward normal of T_1, i.e. $n_F = n_{T_1} = -n_{T_2}$. We note that the normal flux u_n changes its sign when switching from T_1 to T_2,

$$u_{n_{T_1}} = -u_{n_{T_2}},$$

as then also the outer normal changes its direction. If suitable, we also equip $\mu_h \in \Lambda_h^k$ with an index for the direction of the normal vector, such that

$$\mu_h = \mu_{h,n_{T_1}} = -\mu_{h,n_{T_2}}.$$

This means, we identify the space Λ_h^k with the space of facet-wise two-valued functions, where the two values are equal up to their sign.

Note that the space Λ_h^k has exactly as many linearly independent basis functions associated to each facet, as there are degrees of freedom for the normal-normal continuous space Σ_h^k. It is therefore well-suited as space for the Lagrangian multipliers: we observe that the condition "$\sigma_{h,nn}$ is continuous and vanishes on Γ_N" for $\sigma_h \in \widetilde{\Sigma}_h^k$ is *equivalent* to

$$0 = \sum_{F \in \mathcal{F}_h} \int_F [\![\sigma_{h,nn}]\!] \mu_h \, ds = \sum_{T \in \mathcal{T}_h} \int_{\partial T} \sigma_{h,nn} \mu_{h,n} \, ds \qquad \forall \mu_h \in \Lambda_h^k. \tag{5.6}$$

Reordering the facet terms, we obtain an equivalent to Problem 4.11.

Problem 5.1. *Find* $(\sigma_h, u_h, \lambda_h) \in \widetilde{\Sigma}_h^k \times V_h^k \times \Lambda_h^k$ *such that*

$$\begin{aligned} a(\sigma_h, \tau_h) \quad &+ \quad b(\tau_h; u_h, \lambda_h) \quad = \quad \langle F_1, \tau_h \rangle_\Sigma \qquad &&\forall \tau_h \in \widetilde{\Sigma}_h^k, \\ b(\sigma_h; v_h, \mu_h) \quad & \quad = \quad \langle F_2, v_h \rangle_V \qquad &&\forall v_h \in V_h^k, \mu_h \in \Lambda_h^k. \end{aligned} \tag{5.7}$$

Here, $a(\cdot, \cdot), F_1$ *and* F_2 *were defined previously in equations (3.37), (3.39) and (3.40), whereas* $b : \widetilde{\Sigma}_h^k \times (V_h^k \times \Lambda_h^k) \to \mathbb{R}$ *is given by*

$$b(\tau_h; v_h, \mu_h) := \langle \mathrm{div}\, \tau_h, v_h \rangle_V - \sum_{T \in \mathcal{T}_h} \int_{\partial T} \tau_{h,nn} \mu_{h,n} \, ds \tag{5.8}$$

$$= \sum_{T \in \mathcal{T}_h} \left[-\int_T \tau_h : \varepsilon(v_h) \, dx + \int_{\partial T} \tau_{h,nn}(v_{h,n} - \mu_{h,n}) \, ds \right]. \tag{5.9}$$

We immediately see that the original mixed Problem 4.11 and the hybridized version Problem 5.1 are equivalent: for any solution $(\sigma_h, u_h, \lambda_h) \in \widetilde{\Sigma}_h^k \times V_h^k \times \Lambda_h^k$, equation (5.6) ensures that σ_h is normal-normal continuous and therefore $\sigma_h \in \Sigma_h^k$. The second equation in system (5.7) reduces to the original equation $b(\sigma_h, v_h) = \langle F_2, v_h \rangle$ from system (4.3). When testing the first line for conforming $\tau_h \in \Sigma_h^k$, we see, that again the facet terms containing λ_h cancel out, and we are left with the first equation of the original system (4.3). Thus, (σ_h, u_h) is a solution to Problem 4.11.

We observe that $\lambda_{h,n}$ approximates the normal displacement u_n: after element-wise reordering, the first equation of system (5.7) reads

$$\sum_{T \in \mathcal{T}_h} \left[\int_T (A\sigma_h - \varepsilon(u_h)) : \tau_h \, dx + \int_{\partial T} (u_{h,n} - \lambda_{h,n}) \tau_{h,nn} \, ds \right] = 0.$$

There the first term corresponds to an element-wise weak enforcement of Hooke's law, while the second term corresponds to a facet-wise weak identification of $u_{h,n}$ and $\lambda_{h,n}$. Similarly, after a facet-wise reordering of the surface integrals, we see from the second equation in (5.7), that

$$\sum_{T \in \mathcal{T}_h} \int_T (\mathrm{div}\, \sigma_h + f) \cdot v_h \, dx - \sum_{F \in \mathcal{F}_h} \int_F \left([\![\sigma_{h,n\tau}]\!] \cdot v_{h,\tau} + [\![\sigma_{h,nn}]\!] \mu_h \right) ds = 0.$$

Here, the first term corresponds to the equilibrium equation. The latter sum ensures a weak enforcement of continuity of $\sigma_{n\tau}$, and continuity of $\sigma_{h,nn}$, due to the matching number of facet-based degrees of freedom for the stresses and for the Lagrangian multipliers.

5.1.3 Analysis of the hybrid problem

We derive properties of the hybrid method, such as existence and uniqueness of a solution, and a priori error estimates, using the analysis provided for the original formulation of the TD-NNS method in Chapter 4.

We propose to use the following discrete norms for the analysis of the hybrid problem

$$\|\tau_h\|_{\widetilde{\Sigma}_h} := \|\tau_h\|_\Omega, \tag{5.10}$$

$$\|v_h, \mu_h\|_{V_h \times \Lambda_h} := \left(\sum_{T \in \mathcal{T}_h} \left[\|\varepsilon(v_h)\|_T^2 + h_T^{-1} \|v_{h,n} - \mu_{h,n}\|_{\partial T}^2 \right] \right)^{1/2}. \tag{5.11}$$

These norms are directly induced by $\|.\|_{\Sigma_h}$ and $\|.\|_{V_h}$, as were introduced for the mixed problem (see equations (4.27), (4.28)). In this definition, we assume that the triangulation is shape-regular, and thus $h_F \simeq h_T$. Otherwise, the surface terms $h_T^{-1}\|v_{h,n} - \mu_{h,n}\|_{\partial T}^2$ have to be split first element- and then facet-wise, such that the correct factor h_F^{-1} can be used for each facet. Note that $\|.\|_{V_h \times \Lambda_h}$ is really a norm on $V_h^k \times \Lambda_h^k$: if, for some $(v_h, \mu_h) \in V_h^k \times \Lambda_h^k$, we have $\|v_h, \mu_h\|_{V_h \times \Lambda_h} = 0$, we can conclude that

- v_h is continuous, i.e. $v_h \in H^1(\Omega)$
- $v_{h,n} = \mu_{h,n}$, thus $v_{h,n} = 0$ on Γ_D and
- $\varepsilon(v_h) = 0$ in Ω.

Korn's and Friedrichs' inequalities (Theorem 3.14 and Theorem 3.16) ensure then $v_h = 0$, which then implies $\mu_h = 0$. One can even show that, for $v_h \in V_h^k$,

$$\frac{1}{2}\|v_h\|_{V_h}^2 \leq \inf_{\mu_h \in \Lambda_h^k} \|v_h, \mu_h\|_{V_h \times \Lambda_h}^2 \leq \|v_h\|_{V_h}^2.$$

Similar to Lemma 4.13, we can show that the norm for the displacement quantities is equivalent to the usual, broken H^1 norm. We also observe that it is sufficient to use only the piecewise linear projection of the jump of the normal component.

Lemma 5.2. *Let $\mathcal{T}_h$ be a shape-regular, quasi-uniform triangulation of the domain Ω. We have the following norm equivalences on $V_h^k \times \Lambda_h^k$:*

$$\|v_h, \mu_h\|_{V_h \times \Lambda_h}^2 \simeq \sum_{T \in \mathcal{T}_h} \left[\|\nabla v_h\|_T^2 + h_T^{-1}\|v_{h,n} - \mu_{h,n}\|_{\partial T}^2 \right], \tag{5.12}$$

$$\|v_h, \mu_h\|_{V_h \times \Lambda_h}^2 \simeq \sum_{T \in \mathcal{T}_h} \left[\|\varepsilon(v_h)\|_T^2 + h_T^{-1}\|\Pi^1(v_{h,n} - \mu_{h,n})\|_{\partial T}^2 \right], \tag{5.13}$$

where Π^1 denotes the facet-wise projection onto $P^1(\mathcal{F}_h)$. The constants of equivalence do not depend on the mesh parameter h.

Proof. Follows directly from Lemma 4.13. □

Lemma 5.3. *Let $\mathcal{T}_h$ be a shape-regular, quasi-uniform triangulation of the domain Ω. Problem 5.1 is well-posed, there exists a unique solution $(\sigma_h, u_h, \lambda_h) \in \tilde{\Sigma}_h^k \times V_h^k \times \Lambda_h^k$.*

Proof. We apply Brezzi's theory once more, and prove the conditions of Assumption 4.2 in the hybridized setting. We observe that all estimates work along the same lines as in the proofs of Lemma 4.15, Lemma 4.16 and Lemma 4.17. The only difference is that the jump term $[\![v_h]\!]_n$ across a facet is replaced by the two-sided difference $v_{h,n_{T_i}} - \mu_{h,n_{T_i}}$, $i = 1, 2$ on the boundaries of the neighboring elements T_1, T_2. □

In Chapter 4, we provided a-priori bounds for the error in the stress and displacement fields. These estimates can be transferred directly to the hybrid setup, as the discrete solutions coincide. Moreover, we state a bound for the error in the Lagrangian multiplier.

Lemma 5.4. *Let $\mathcal{T}_h$ be a shape-regular, quasi-uniform triangulation of the domain Ω. Let moreover $(\sigma_h, u_h, \lambda_h) \in \tilde{\Sigma}_h^k \times V_h^k \times \Lambda_h^k$ be the solution to the hybrid problem (Problem 5.1), and let (σ, u) be the solution to the classical problem (Problem 3.1). Assuming (σ, u) lies in $H_{SYM}^m(\Omega) \times [H^{m+1}(\Omega)]^d$ for integer $1 \leq m \leq k$, the discretization error is bounded by*

$$\|\sigma - \sigma_h\|_{\Sigma_h} + \|u - u_h\|_{V_h} \leq ch^m \left(|\sigma|_{H^m(\Omega)} + |u|_{H^{m+1}(\Omega)} \right) \tag{5.14}$$

$$\left(\sum_{T \in \mathcal{T}_h} h_T^{-1} \|u_n - \lambda_{h,n}\|_{\partial T}^2 \right)^{1/2} \leq ch^m \left(|\sigma|_{H^m(\Omega)} + |u|_{H^{m+1}(\Omega)} \right). \tag{5.15}$$

Proof. The proof relies on techniques from [AB85], where a similar estimate was shown for the hybridized, mixed formulation of Poisson's equation. The estimate (5.14) for σ_h and u_h follows directly from the error estimate in the mixed setting (Theorem 4.25), as the discrete solutions coincide. Thus, we are left with proving the error bound (5.15) for the Lagrangian multiplier. We treat this estimate facet-wise. Let $T \in \mathcal{T}_h$ be an arbitrary element, and $F \in \mathcal{F}(T)$ one of its facets. Let Π_F^k be the L^2 projection onto $P^k(F)$, then we may split the discretization error

$$\|u_n - \lambda_{h,n}\|_F \leq \|u_n - \Pi_F^k u_n\|_F + \|\lambda_{h,n} - \Pi_F^k u_n\|_F. \tag{5.16}$$

The approximation error on the facet is bounded by standard error estimates,

$$\|u_n - \Pi_F^k u_n\|_F \leq ch^{m+1/2} |u|_{H^{m+1}}. \tag{5.17}$$

To estimate the contribution of the second term in (5.16), we construct a special test function $\tau_h \in \tilde{\Sigma}_h^k$ to use in the discrete hybrid variational formulation (5.7). There exists a tensor-valued finite element function $\tau_h^{F,T} \in \Sigma^k(T)$ with support only on element T satisfying

$$\tau_{h,nn}^{F,T} = \lambda_{h,n_T} - \Pi_F^k u_{n_T} \quad \text{on } F,$$
$$\tau_{h,nn}^{F,T} = 0 \quad \text{on } \tilde{F} \in \mathcal{F}(T), \tilde{F} \neq F.$$

A construction of this function can be done using the facet space $\Sigma_h^{k,f}(T)$, as done in the proof of Lemma 4.17. This tensor is bounded by

$$\left\|\tau_h^{F,T}\right\|_T \leq c h_T^{1/2} \left\|\lambda_{h,n} - \Pi_F^k u_n\right\|_F, \tag{5.18}$$

which follows from the equivalence of norms shown in Lemma 4.14. We build the desired test function

$$\tau_h := \sum_{T \in \mathcal{T}_h} \sum_{F \in \mathcal{F}(T)} \tau_h^{F,T} \in \widetilde{\Sigma}_h^k.$$

For this τ_h, the first line of system (5.7) reads

$$a(\sigma_h, \tau_h) + \langle \operatorname{div} \tau_h, u_h \rangle_V - \sum_{T \in \mathcal{T}_h} \sum_{F \in \mathcal{F}(T)} \int_F (\lambda_{h,n} - \Pi_F^k u_n) \lambda_{h,n}\, ds = \langle F_1, \tau_h \rangle_\Sigma. \tag{5.19}$$

For the smooth solution (σ, u), we can deduce that

$$a(\sigma, \tau_h) + \langle \operatorname{div} \tau_h, u \rangle_V - \sum_{T \in \mathcal{T}_h} \sum_{F \in \mathcal{F}(T)} \int_F (\lambda_{h,n} - \Pi_F^k u_n) u_n\, ds = \langle F_1, \tau_h \rangle_\Sigma. \tag{5.20}$$

using Green's formula element-wise. Subtracting equation (5.19) from (5.20), yields the equivalence

$$\sum_{T \in \mathcal{T}_h} \sum_{F \in \mathcal{F}(T)} \left\|\lambda_{h,n} - \Pi_F^k u_n\right\|_F^2 = -a(\sigma - \sigma_h, \tau_h) - \langle \operatorname{div} \tau_h, u - u_h \rangle_V.$$

There we used the defining property of the orthogonal projection on facet F,

$$\left\|\lambda_{h,n} - \Pi_F^k u_n\right\|_F^2 = \int_F \left(\lambda_{h,n} - \Pi_F^k u_n\right)\left(u_n - \lambda_{h,n}\right) ds.$$

Continuity of $a(\cdot,\cdot)$ and the divergence operator in the discrete setting together with (5.18) then ensure

$$\sum_{T \in \mathcal{T}_h} h_T^{-1/2} \sum_{F \in \mathcal{F}(T)} \left\|\lambda_{h,n} - \Pi_F^k u_n\right\|_F \leq c \left(\|\sigma - \sigma_h\|_{\Sigma_h} + \|u - u_h\|_{V_h}\right).$$

This concludes the proof, as we already bounded the discretization error for the stresses and displacements according to (5.14). □

Similarly to Corollary 4.18, we see that the order associated to the facets can be reduced for the stress and Lagrange parameter space.

Corollary 5.5. *The results of Lemma 5.3 and Lemma 5.4 still hold true when using the approximation spaces $\widetilde{\Sigma}_h^{k,k_f} \times V_h^k \times \Lambda_h^{k_f}$, where $k_f := \max(1, k-1)$. There, $\widetilde{\Sigma}_h^{k,k_f}$ is defined as the discontinuous equivalent of Σ_h^{k,k_f} of overall order k, where the polynomial order of the facet contributions is set to k_f.*

So far, we have seen that the hybridized problem Problem 5.1 has the same properties as the original mixed TD-NNS formulation in Problem 4.11. One of the main benefits of the hybridization process is the fact that, when the stress unknowns are eliminated locally, one ends up with a symmetric, positive definite system matrix. As all degrees of freedom stemming from the stresses are local to an element, this elimination process can be done totally local, and is of globally optimal computational complexity. Throughout the following, we denote A_h, B_h as the linear operators induced by the bilinear forms $a(\cdot, \cdot)$ and $b(\cdot, \cdot)$ on the discrete spaces $\widetilde{\Sigma}_h^k$ and $V_h^k \times \Lambda_h^k$,

$$\langle A_h \sigma_h, \tau_h \rangle = a(\sigma_h, \tau_h) \qquad \text{for } \sigma_h, \tau_h \in \widetilde{\Sigma}_h^k,$$
$$\langle B_h \tau_h; v_h, \mu_h \rangle = b(\tau_h; v_h, \mu_h) \qquad \text{for } \tau_h \in \widetilde{\Sigma}_h^k, \ v_h \in V_h^k, \ \mu_h \in \Lambda_h^k.$$

The Schur complement operator S_h is defined as

$$S_h := B_h A_h^{-1} B_h^*. \tag{5.21}$$

The Schur operator induces then again a bilinear form $s(\cdot, \cdot)$ on the discrete space $V_h^k \times \Lambda_h^k$ via

$$s(u_h, \lambda_h; v_h, \mu_h) := \langle S_h(u_h, \lambda_h); v_h, \mu_h \rangle \qquad \text{for } (u_h, \lambda_h), (v_h, \mu_h) \in V_h^k \times \Lambda_h^k.$$

The following Schur complement problem is equivalent to Problems 4.11 and 5.1.

Problem 5.6. *Find* $u_h \in V_h^k$, $\lambda_h \in \Lambda_h^k$ *such that*

$$s(u_h, \lambda_h; v_h, \mu_h) = H_h(v_h, \mu_h) \qquad \forall v_h \in V_h^k, \mu_h \in \Lambda_h^k, \tag{5.22}$$

where the Schur bilinear-form stems from the Schur complement defined in (5.21), and the right hand side $H_h \in (V_h^k \times \Lambda_h^k)^*$ *is given by*

$$H_h := B_h A_h^{-1} F_1 - F_2.$$

The stress field $\sigma_h \in \widetilde{\Sigma}_h^k$ *is determined uniquely by*

$$a(\sigma_h, \tau_h) = F_1(\tau_h) - b(\tau_h; u_h, \lambda_h) \qquad \forall \tau_h \in \widetilde{\Sigma}_h^k.$$

We note that, due to the stability analysis provided for the mixed system (5.7), the Schur complement operator is bounded and coercive with respect to the norm $\|\cdot\|_{V_h \times \Lambda_h}$, i.e. there exist constants $\tilde{c}_{s,1}, \tilde{c}_{s,2} > 0$ such that

$$\tilde{c}_{s,1} \|v_h, \mu_h\|_{V_h \times \Lambda_h}^2 \leq s_h(v_h, \mu_h; v_h, \mu_h) \leq \tilde{c}_{s,2} \|v_h, \mu_h\|_{V_h \times \Lambda_h}^2 \qquad \text{for all } v_h \in V_h^k, \mu_h \in \Lambda_h^k. \tag{5.23}$$

5.1.4 Implementation issues

We shortly outline how the implementation of the TD-NNS method can be simplified by the hybridization procedure. We discuss a possible nodal basis for the Lagrange parameter space Λ_h^k, and a realization of the finite element basis functions.

5.1.4.1 Finite elements for the discontinuous stress space

In Section 4.4, we provided basis functions for triangular and tetrahedral elements of arbitrary order for the stress space Σ_h^k. In the hybridized setting, the global finite element basis functions do not need to satisfy normal-normal continuity. Therefore, one can use any basis of $P^k(T)$ for the local space $\widetilde{\Sigma}^k(T)$. A possible choice are the bases $\{q_{ij} : 0 \leq i+j \leq k\}, \{q_{ijl} : 0 \leq i+j+l \leq k\}$ given in Section 4.4.1 in two or three space dimensions respectively. However, when one wants to restrict the method to the subspace Σ_h^{k,k_f} as proposed in Corollary 4.18, one has to keep the basis Ψ_k^Σ proposed for the conforming method. The degrees of freedom living on a facet are then split, such that they are then associated to the interiors of the two adjacent elements.

5.1.4.2 Finite elements for the Lagrangian multiplier space

Finite elements for the Lagrange parameters are not assigned to elements in the mesh, but to facets. We propose to use the triplet $(F, \Lambda^k(F), \mathcal{N}_k^\Lambda(F))$ on facet $F \in \mathcal{F}_h$. There, we define the local space $\Lambda^k(F) := P^k(F)$. Given a basis $\{q_i\}$ for $P^k(F)$, the set of nodal values corresponding to F is $\mathcal{N}_k^\Lambda(F) = \{N_{F,i}^\Lambda\}$, where

$$N_{F,i}^\Lambda(\mu) := \int_F \mu \, q_i \, ds. \tag{5.24}$$

On the reference facet $\hat{F}$, we now propose a specific basis $\Psi_{\hat{F},k}^\Lambda$, such that $\Lambda^k(\hat{F}) = \operatorname{span} \Psi_{\hat{F},k}^\Lambda$. In two space dimensions, let $\hat{F} = [V_\alpha, V_\beta]$ be the oriented reference segment between vertices V_α, V_β, as described in Section 4.4. We define the basis $\Psi_{\hat{F},k}^\Lambda = \{\psi_{\hat{F},i}^\Lambda : i = 0, \ldots, k\}$, where

$$\psi_{\hat{F},i}^\Lambda := q_i(\lambda_\alpha, \lambda_\beta).$$

In three space dimensions, we need to provide a basis on the oriented triangular facet $\hat{F} = [V_\alpha, V_\beta, V_\gamma]$. We set $\Psi_{\hat{F},k}^\Lambda = \{\psi_{\hat{F},ij}^\Lambda : 0 \leq i+j \leq k\}$, where

$$\psi_{\hat{F},ij}^\Lambda := q_{ij}(\lambda_\alpha, \lambda_\beta, \lambda_\gamma).$$

Here, the polynomials q_i, q_{ij} are taken from Section 4.4.1, equations (4.41) and (4.42). Note that, due to its construction, the basis is orthogonal in the L^2 sense, both in two and three space dimensions.

The functions are transformed to any facet $F \in \mathcal{F}_h$ without further computational effort. For $x \in T$, let $\hat{x} \in \hat{T}$ be the corresponding point on the reference element; we transform $\hat{\mu}_h$ on the reference element to $\mu_h := \Phi_F^\Lambda \hat{\mu}_h$, where

$$\left(\Phi_F^\Lambda \hat{\mu}_h\right)(x) := \hat{\mu}(\hat{x}).$$

The global finite element space is composed from the local ones without further continuity assumptions:

$$\Lambda_h^k = \bigoplus_{F \in \mathcal{F}_h} \Lambda_h^k(F).$$

5.1.4.3 A decomposition for the Lagrange parameter space

We decomposed the Nédélec space V_h^k into a global, low-order space $V_{h,0}$ and local, high-order spaces $\tilde{V}_X^k$ for edges, facets and elements $X \in \mathcal{X}_h$. We similarly do so for the Lagrange parameter space Λ_h^k. Here, the low-order space $\Lambda_{h,0}$ shall consist of the piecewise linear contributions. The high-order spaces $\tilde{\Lambda}_F^k$ each correspond to a facet $F \in \mathcal{F}_h$, and contain the facet basis functions of orders two and higher. Precisely, we set

$$\Lambda_{h,0} := \text{span}\Big(\bigcup_{F \in \mathcal{F}_h} \{\psi_F^\Lambda \in \Psi_{F,k}^\Lambda : \deg(\psi_F^\Lambda) \leq 1\}\Big), \tag{5.25}$$

$$\tilde{\Lambda}^k := \bigoplus_{F \in \mathcal{F}_h} \tilde{\Lambda}_F^k, \tag{5.26}$$

$$\tilde{\Lambda}_F^k := \text{span}\{\psi_F^\Lambda \in \Psi_{F,k}^\Lambda : \deg(\psi_F^\Lambda) \geq 2\}. \tag{5.27}$$

Due to the construction of the basis, this decomposition is orthogonal in the L^2 sense. We will heavily use this fact when treating almost incompressible materials.

5.2 Preconditioning

In this section, we propose a suitable preconditioner C_h^{-1} for the Schur complement equation (5.22). We define and analyze an additive Schwarz block preconditioner, where the finite element spaces are split into a global, low-order part, and local high-order contributions associated to one edge, facet or element. We show that its inverse C_h is spectrally equivalent to S_h, where the constants of equivalence do not depend on the mesh size h,

$$\langle S_h(v_h, \mu_h); v_h, \mu_h \rangle \simeq \langle C_h(v_h, \mu_h); v_h, \mu_h \rangle \qquad \forall v_h \in V_h^k, \mu_h \in \Lambda_h^k.$$

To do so, we first introduce the notion of additive Schwarz preconditioners formally in the subsequent section.

5.2.1 Additive Schwarz block preconditioners

The theory of additive Schwarz preconditioners goes back to the historic work of [Sch70]. To define the preconditioner, the finite element spaces are divided in sub-blocks, which may be overlapping. Well-known examples of additive Schwarz preconditioners are standard multilevel preconditioners, or preconditioners stemming from domain decomposition (DD) methods. For the multilevel methods, the sub-blocks usually correspond to finite element spaces on coarser grids. In DD, one considers a splitting of the finite element space, which is induced by a splitting of the domain in, possibly overlapping, patches. Concerning theory on multilevel and multigrid methods, we refer to [Hac85, BPX90, Bra93, BZ00], where basic techniques also used in our approach are derived. Multigrid methods have also been developed for the dual mixed Hellinger-Reissner formulation, we refer to [PW06], where techniques for $H(\text{div})$ conforming problems developed in [AFW97b, AFW98, AFW00] are applied. In [Sch99, Wie00], multigrid methods for the displacement-pressure formulation of elasticity are provided. For an extensive introduction on DD methods, we point the reader to the monograph [TW05]. For high-order methods, it is common to see each edge, facet or element, or patches consisting of several such entities, as subdomain. Theory for p- and hp-methods for elliptic, H^1 conforming problems can be found e.g. in [Man90, Ain96a, Ain96b, SMPZ08]. For problems in $H(\text{curl})$, as the different formulations of Maxwell's equations, we refer to [AFW00, HT00, PZ02].

In our approach, we divide the system matrix into sub-blocks, where one block corresponds to the low-order degrees of freedom, and is global. The high-order nodal values are organized into face, edge and element blocks, and are local. The preconditioner then consists of the sum of the inverses of the respective sub-blocks.

We now introduce the notion of an additive Schwarz block preconditioner formally.

Definition 5.7. *Let S_h be a symmetric, positive definite operator on the finite element space Q_h. Let moreover $\{Q_i : i = 1, \ldots, s\}$ be a not necessarily unique splitting of Q_h, such that $Q_h = \sum_{i=1}^{s} Q_i$. Let $E_i : Q_h \to Q_i$ be corresponding restriction operators. Then the additive Schwarz preconditioner C_h^{-1} for S_h is defined by its application via*

$$C_h^{-1} q := \sum_{i=1}^{s} E_i^* C_i^{-1} E_i q, \qquad \text{where } C_i = E_i S_h E_i^*. \tag{5.28}$$

There, C_i^{-1} is the inverse of C_i on the subspace Q_i.

5.2.2 A preconditioner for hybridized elasticity

We define an additive Schwarz block preconditioner for the Schur complement equation in Problem 5.6. We decompose the finite element spaces V_h^k, Λ_h^k, as was proposed in Sections 4.2.2.6 and 5.1.4.3, namely

$$V_h^k = V_{h,0} + \sum_{X \in \mathcal{X}_h} \tilde{V}_X^k, \qquad \Lambda_h^k = \Lambda_{h,0} + \sum_{F \in \mathcal{F}_h} \tilde{\Lambda}_F^k. \tag{5.29}$$

Remember $\mathcal{X}_h = \mathcal{E}_h \cup \mathcal{F}_h \cup \mathcal{T}_h$ is the set of all edges, facets and elements in the mesh. There, we set $V_{h,0}$ as the lowest-order Nédélec type I space, and $\Lambda_{h,0}$ as the space containing the piecewise linear facet fields. The high-order spaces are local; for an edge, facet or element $X \in \mathcal{X}_h$, the space $\tilde{V}_X^k$ is spanned by the high-order shape basis functions connected to X, and has support on only a few elements surrounding X. Similarly, for $F \in \mathcal{F}_h$, the space $\tilde{\Lambda}_F^k$ consists of the shape basis functions of degree two and higher on facet F. The supports of the different high-order subspaces are even disjoint. Any $v_h \in V_h^k, \mu_h^k \in \Lambda_h^k$ can be decomposed uniquely such that

$$v_h = v_0 + \sum_{X \in \mathcal{X}_h} v_X, \qquad \text{where } v_0 \in V_{h,0}, \ v_X \in \tilde{V}_X^k, \qquad (5.30)$$

$$\mu_h = \mu_0 + \sum_{F \in \mathcal{F}_h} \mu_F, \qquad \text{where } \mu_0 \in \Lambda_{h,0}, \ \mu_F \in \tilde{\Lambda}_F^k. \qquad (5.31)$$

Let, for an edge, facet or element $X \in \mathcal{X}_h$ and a facet $F \in \mathcal{F}_h$

$$E_0 : V_h^k \times \Lambda_h^k \to V_{h,0} \times \Lambda_{h,0}, \qquad E_X^V : V_h^k \times \Lambda_h^k \to \tilde{V}_X^k \times \{0\}, \qquad E_F^\Lambda : V_h^k \times \Lambda_h^k \to \{0\} \times \tilde{\Lambda}_F^k \quad (5.32)$$

be the respective restriction operators for the finite element subspaces. There the operator E_0 is consists of the nodal Nédélec interpolation operator, and an L^2 projection onto the piecewise linear functions for Λ_h^k. Note that we thereby put the low-order subspaces $V_{h,0}$ and $\Lambda_{h,0}$ into one sub-block. The different blocks of the Schur operator S_h with respect to this splitting read for $X \in \mathcal{X}_h, F \in \mathcal{F}_h$

$$S_0 := E_0^* S_h E_0, \qquad S_X^V := \left(E_X^V\right)^* S_h E_X^V, \qquad S_F^\Lambda := \left(E_F^\Lambda\right)^* S_h E_F^\Lambda. \qquad (5.33)$$

Using the splittings (5.30) and (5.31) for $v_h \in V_h^k, \mu_h \in \Lambda_h^k$, the preconditioner C_h^{-1} is defined by

$$C_h^{-1}(v_h, \mu_h) := S_0^{-1}(v_0, \mu_0) + \sum_{X \in \mathcal{X}_h} \left(S_X^V\right)^{-1}(v_X, 0) + \sum_{F \in \mathcal{F}_h} \left(S_F^\Lambda\right)^{-1}(0, \mu_F), \qquad (5.34)$$

where all inverses are defined on the respective subspaces.

We state the main theorem of this section, which assures that C_h is spectrally equivalent to the Schur complement operator S_h.

Theorem 5.8. *For a domain $\Omega \subset \mathbb{R}^d, d = 2, 3$ satisfying Assumption 3.2, let $\mathcal{T}_h$ be a shape-regular, quasi-uniform triangulation. Let S_h denote the Schur complement operator from Problem 5.6, and C_h^{-1} be the corresponding additive Schwarz block preconditioner defined by equation (5.34). Then S_h and C_h are spectrally equivalent,*

$$\langle S_h(v_h, \mu_h); v_h, \mu_h \rangle \simeq \langle C_h(v_h, \mu_h); v_h, \mu_h \rangle \qquad \forall v_h \in V_h^k, \mu_h \in \Lambda_h^k. \qquad (5.35)$$

The constants of equivalence do not depend on the mesh size h.

Proof. Postponed to Section 5.2.3. □

5.2.3 Condition number estimates

In the following, we proof two lemmas, which allow us to bound the different components (v_0, μ_0), $(v_X, 0)$ and $(0, \mu_F)$ by (v_h, μ_h) in the discrete norm $\|\cdot\|_{V_h \times \Lambda_h}$. Together with an estimate, which allows to bound (v_h, μ_h) from above by the different contributions, they are employed in the subsequent proof of Theorem 5.8.

In the following, we assume that the conditions from Theorem 5.8 hold, i.e that Ω satisfies Assumption 3.2, its triangulation $\mathcal{T}_h$ is simplicial, quasi-uniform, and shape-regular, and that S_h and C_h^{-1} are the Schur complement operator and corresponding additive Schwarz block preconditioner.

Lemma 5.9. *For $(v_h, \mu_h) \in V_h^k \times \Lambda_h^k$, let $(v_0, \mu_0) \in V_{h,0} \times \Lambda_{h,0}$ be the projection onto the low-order spaces. Then there exists some $c > 0$ independent of the mesh size h, such that*

$$\|v_0, \mu_0\|_{V_h \times \Lambda_h} \leq c \|v_h, \mu_h\|_{V_h \times \Lambda_h}.$$

Proof. Let $v_h \in V_h^k, \mu_h \in \Lambda_h^k$ be arbitrary, and v_0, μ_0 their respective low-order components. Moreover, $\tilde{v}_h := v_h - v_0$ and $\tilde{\mu}_h := \mu_h - \mu_0$ shall denote the high-order contributions. In the following estimate, we use that $\varepsilon(v_0) = 0$ to obtain the second equality, and that the operator norm of the L^2 projection is less or equal to one as well as the triangle inequality in the third line.

$$\begin{aligned}
\|v_0, \mu_0\|_{V_h \times \Lambda_h}^2 &= \sum_{T \in \mathcal{T}_h} \left[\|\varepsilon(v_0)\|_T^2 + h_T^{-1} \|v_{0,n} - \mu_{0,n}\|_{\partial T}^2 \right] \\
&= \sum_{T \in \mathcal{T}_h} h_T^{-1} \left\| \Pi^1 \left((\mathcal{I}_{h,0}^V v_h)_n - \mu_{h,n} \right) \right\|_{\partial T}^2 \\
&\leq \sum_{T \in \mathcal{T}_h} 2 h_T^{-1} \left(\| \underbrace{\left(\mathcal{I}_{h,0}^V v_h - v_h \right)}_{=:\tilde{v}_h \in \tilde{V}_h^k}{}_n \|_{\partial T}^2 + \|v_{h,n} - \mu_{h,n}\|_{\partial T}^2 \right).
\end{aligned}$$

Corollary 4.29 (piecewise Korn inequality for the high order space $\tilde{V}^k$) now ensures

$$\|v_0, \mu_0\|_{V_h \times \Lambda_h}^2 \leq c \sum_{T \in \mathcal{T}_h} \left[\|\varepsilon(\tilde{v}_h)\|_T^2 + h_T^{-1} \|v_{h,n} - \mu_{h,n}\|_{\partial T}^2 \right].$$

As $\varepsilon(\tilde{v}_h) = \varepsilon(v_h)$, this is the required result. □

Lemma 5.10. *For $(v_h, \mu_h) \in V_h^k \times \Lambda_h^k$, let $v_X \in \tilde{V}_X^k$ and $\mu_F \in \tilde{\Lambda}_F^k$ be the sub-blocks matching some $X \in \mathcal{X}_h$, $F \in \mathcal{F}_h$ respectively. Then there exists some $c > 0$ independent of h such that*

$$\sum_{X \in \mathcal{X}_h} \|v_X, 0\|_{V_h \times \Lambda_h} + \sum_{F \in \mathcal{F}_h} \|0, \mu_F\|_{V_h \times \Lambda_h} \leq c \|v_h, \mu_h\|_{V_h \times \Lambda_h}.$$

Proof. We split the proof into two parts, one for the estimation of the contributions stemming from v_X, $X \in \mathcal{X}_h$, and one for the respective contributions from μ_F, $F \in \mathcal{F}_h$.

1. For $v_h \in V_h^k$ decomposed according to (5.30), we see, employing Corollary 4.29 to estimate the boundary terms by the strain tensor for the high-order contributions v_X,

$$\sum_{X \in \mathcal{X}_h} \|v_X, 0\|_{V_h \times \Lambda_h} = \sum_{X \in \mathcal{X}_h} \sum_{T \in \Delta_X} \left[\|\varepsilon(v_X)\|_T^2 + h_T^{-1} \|v_{X,n}\|_{\partial T}^2 \right]$$

$$\leq \sum_{X \in \mathcal{X}_h} \sum_{T \in \Delta_X} \|\varepsilon(v_X)\|_T^2 \leq \sum_{T \in \mathcal{T}_h} \|\varepsilon(v_h)\|_T^2.$$

We moreover used that the number of components v_X with support on an element T is bounded.

2. In order to estimate the contributions stemming from μ_h, we once more need the splitting $v_h = v_0 + \tilde{v}_h$ into the low- and high-order contributions. We utilize Lemma 5.9 and Corollary 4.29 to obtain

$$\sum_{F \in \mathcal{F}_h} \|0, \mu_F\|_{V_h \times \Lambda_h} = 2 \sum_{F \in \mathcal{F}_h} h_T^{-1} \|\mu_F\|_F^2$$

$$= \sum_{T \in \mathcal{T}_h} h_T^{-1} \|\mu_{h,n} - \mu_{0,n} - v_{h,n} + v_{0,n} + \tilde{v}_{h,n}\|_{\partial T}^2$$

$$\leq 3 \sum_{T \in \mathcal{T}_h} h_T^{-1} \left(\|\mu_{h,n} - v_{h,n}\|_{\partial T}^2 + \|\mu_{0,n} - v_{0,n}\|_{\partial T}^2 + \|\tilde{v}_{h,n}\|_{\partial T}^2 \right)$$

$$\overset{\text{C. 4.29}}{\leq} c \sum_{T \in \mathcal{T}_h} \left[\|\varepsilon(\tilde{v}_h)\|_T^2 + h_T^{-1} \left(\|\mu_{h,n} - v_{h,n}\|_{\partial T}^2 + \|\mu_{0,n} - v_{0,n}\|_{\partial T}^2 \right) \right]$$

$$\overset{\text{L. 5.9}}{\leq} c \, \|v_h, \mu_h\|_{V_h \times \Lambda_h}.$$

This concludes our proof.

□

Lemma 5.11. *Let (v_h, μ_h) be decomposed according to (5.30), (5.31). Then there exists a constant $c > 0$ independent of the mesh size h such that*

$$\|v_h, \mu_h\|_{V_h \times \Lambda_h}^2 \leq c \left[\|v_0, \mu_0\|_{V_h \times \Lambda_h}^2 + \sum_{X \in \mathcal{X}_h} \|v_X, 0\|_{V_h \times \Lambda_h}^2 + \sum_{F \in \mathcal{F}_h} \|0, \mu_F\|_{V_h \times \Lambda_h}^2 \right].$$

Proof. The triangle inequality ensures

$$\|v_h, \mu_h\|_{V_h \times \Lambda_h} \leq \|v_0, \mu_0\|_{V_h \times \Lambda_h} + \left\| \sum_{X \in \mathcal{X}_h} (v_X, 0) \right\|_{V_h \times \Lambda_h} + \left\| \sum_{F \in \mathcal{F}_h} (0, \mu_F) \right\|_{V_h \times \Lambda_h}.$$

As only a globally limited number of blocks have support on an element T or facet F, and the supports of the respective blocks for the Lagrangian multipliers are even disjoint, one directly obtains the required result.

□

Proof of Theorem 5.8. We show the spectral equivalence of the inverse of the additive Schwarz block preconditioner C_h, and the Schur complement operator S_h. In our case, the different blocks are non-overlapping, thus

$$C_h(v_h, \mu_h) = S_{h,0}(v_0, \mu_0) + \sum_{X \in \mathcal{X}_h} S_{h,X}^V(v_X, 0) + \sum_{F \in \mathcal{F}_h} S_{h,F}^\Lambda(0, \mu_F)$$

for any $v_h \in V_h^k, \mu_h \in \Lambda_h^k$. As the Schur operator S_h is bounded and coercive with respect to the discrete norm $\|\cdot\|_{V_h \times \Lambda_h}$ (see (5.23)), it suffices to show

$$\|v_h, \mu_h\|_{V_h \times \Lambda_h}^2 \simeq \|v_0, \mu_0\|_{V_h \times \Lambda_h}^2 + \sum_{X \in \mathcal{X}_h} \|v_X, 0\|_{V_h \times \Lambda_h}^2 + \sum_{F \in \mathcal{F}_h} \|0, \mu_F\|_{V_h \times \Lambda_h}^2.$$

This follows from Lemmas 5.9, 5.10 and 5.11. □

5.3 Application to nearly incompressible elasticity

In the following, we treat the case of almost incompressible materials. So far, we discussed a hybrid formulation of the TD-NNS method, and its positive implications, when it comes to implementation issues: much simpler elements can be used, and, after local elimination of the stress degrees of freedom, one ends up with a symmetric positive definite system matrix. However, we did not improve the behavior of the method for nearly incompressible materials.

Throughout this section, we will assume that the material is homogenous, isotropic, and linearly elastic. Thus, the compliance tensor is uniquely determined by the Lamé constants $\bar\mu, \bar\lambda$. The bilinear form $a(\cdot, \cdot)$ as defined in (3.37) then reads

$$a(\sigma, \tau) := \int_\Omega \frac{1}{2\bar\mu} \operatorname{dev} \sigma_h : \operatorname{dev} \tau_h + \frac{1}{d\bar\lambda + 2\bar\mu} \operatorname{tr}(\sigma_h) \operatorname{tr}(\tau_h) \, dx. \tag{5.36}$$

We see, that in the incompressible limit of $\bar\lambda \to \infty$, the strain depends only on the deviator of the stress. Due to this, the bilinear form is not coercive on the whole space $\widetilde\Sigma_h^k$. In the following, we stabilize $a(\cdot, \cdot)$ by adding element-wise divergence terms. To stay consistent, we modify the right hand side accordingly. For $\sigma_h, \tau_h \in \widetilde\Sigma_h^k$, we set

$$a^S(\sigma_h, \tau_h) := a(\sigma_h, \tau_h) + \sum_{T \in \mathcal{T}} h_T^2 \int_T \operatorname{div} \sigma_h \cdot \operatorname{div} \tau_h \, dx, \tag{5.37}$$

$$\langle F_1^S, \tau_h \rangle_\Sigma := \langle F_1, \tau_h \rangle_\Sigma - \sum_{T \in \mathcal{T}} h_T^2 \int_T f \cdot \operatorname{div} \tau_h \, dx. \tag{5.38}$$

As the balance equation reads $-\operatorname{div} \sigma = f$, replacing $a(\cdot, \cdot)$ by $a^S(\cdot, \cdot)$ and F_1 by F_1^S in Problem 5.1 yields another consistent approximation of the problem of mixed elasticity.

Problem 5.12. *Given a regular, quasi-uniform triangulation $\mathcal{T}_h$, find $(\sigma_h, u_h, \lambda_h) \in \widetilde{\Sigma}_h^k \times V_h^k \times \Lambda_h^k$ such that*

$$\begin{aligned} a^S(\sigma_h, \tau_h) &+ b(\tau_h; u_h, \lambda_h) = \langle F_1^S, \tau_h \rangle_\Sigma & \forall \tau_h \in \widetilde{\Sigma}_h^k, \\ b(\sigma_h; v_h, \mu_h) &= \langle F_2, v_h \rangle_V & \forall v_h \in V_h^k, \mu_h \in \Lambda_h^k. \end{aligned} \quad (5.39)$$

Here, $a^S(\cdot, \cdot)$, F_1^s are defined by the relations (5.37), (5.38), whereas $b(\cdot, \cdot)$ is given by (5.8), (5.9), and F_2 is defined by (3.40).

When analyzing the stabilized hybrid problem above, we use different, parameter dependent norms. We introduce these norms subsequently. In the end, we show that the additive Schwarz block preconditioner proposed in Section 5.2.2 is also suitable in the almost incompressible case.

5.3.1 Discrete norms

We define broken norms for estimating $\tau_h \in \widetilde{\Sigma}_h^k$ and $(v_h, \mu_h) \in V_h^k \times \Lambda_h^k$. They are similar to the original choices (5.10) and (5.11), but additionally contain the Lamé parameter $\bar{\lambda}$.

$$\|\tau_h\|_{\Sigma_h, S}^2 := \inf_{\bar{p} \in P^0(\mathcal{T}_h)} \|\tau_h - \bar{p}I\|_\Omega^2 + \frac{1}{\bar{\lambda}} \|\bar{p}I\|_\Omega^2 + \sum_{T \in \mathcal{T}} h_T^2 \|\operatorname{div} \tau_h\|_T^2, \quad (5.40)$$

$$\|v_h, \mu_h\|_{V_h \times \Lambda_h, S}^2 := \sum_{T \in \mathcal{T}_h} \left[\|\varepsilon(v_h)\|_T^2 + h_T^{-1} \|v_{h,n} - \mu_{h,n}\|_{\partial T}^2 + \frac{\bar{\lambda}}{|T|} \left| \int_T \mu_{h,n} ds \right|^2 \right]. \quad (5.41)$$

For the stresses, the first term corresponds to an element-wise deviatoric part of τ_h. In the limiting case of $\bar{\lambda} = \infty$, only this deviatoric part is taken into account. The scaling of h_T in the last term is chosen appropriately to match with the L^2 norm. We note that we added the same kind of element-wise divergence to the bilinear form $a(\cdot, \cdot)$ to stabilize the hybridized system.

For the displacements, we already encountered the first two terms in the definition of $\|.\|_{V_h \times \Lambda_h}$ in the broken norm (5.11) for the hybrid formulation. The last term $\int_T \mu_{h,n} ds$ reflects the change of volume of an element T. As the material becomes incompressible, i.e. $\bar{\lambda} \to \infty$, such changes are penalized.

Before we come to stability analysis, we take a closer look to the parameter-dependent norm defined for the stress field. On the triangulation $\mathcal{T}_h$, we can divide any $\tau_h \in \widetilde{\Sigma}_h^k$ into a deviatoric and a trace part:

$$\tau_h = \operatorname{dev}_T \tau_h + \tfrac{1}{d} \operatorname{tr}_T(\tau_h) I.$$

There, $\operatorname{dev}_T \tau_h$ and $\operatorname{tr}_T(\tau_h)$ are defined element-wise; we set for $T \in \mathcal{T}_h$

$$\operatorname{tr}_T(\tau_h)\big|_T := \frac{1}{|T|} \int_T \operatorname{tr}(\tau_h) ds, \qquad \operatorname{dev}_T \tau_h\big|_T := \tau_h - \tfrac{1}{d} \operatorname{tr}_T(\tau_h) I. \quad (5.42)$$

The trace part $\operatorname{tr}_T(\tau_h)$ is therefore the element-wise average of the trace of τ_h. One can easily see that, independently of $\bar{\lambda} > 0$,

$$\|\operatorname{dev}_T \tau_h\|_\Omega^2 = \inf_{\bar{p} \in P^0(\mathcal{T}_h)} \|\tau_h - \bar{p}I\|_\Omega^2 + \frac{1}{\bar{\lambda}}\|\bar{p}I\|_\Omega^2.$$

Thus, for any $\tau_h \in \widetilde{\Sigma}_h$, we can evaluate $\|\operatorname{dev}_T \tau_h\|_{\widetilde{\Sigma}_h, S}$ by

$$\|\operatorname{dev}_T \tau_h\|_{\widetilde{\Sigma}_h, S}^2 = \|\operatorname{dev}_T \tau_h\|_\Omega^2 + \sum_{T \in \mathcal{T}_h} h_T^2 \|\operatorname{div} \tau_h\|_T^2. \tag{5.43}$$

This follows trivially, as the element-wise divergence of the constant trace part $\operatorname{tr}_T(\tau_h)I$ vanishes. On the other hand, we see for a piecewise constant function $q \in P^0(\mathcal{T}_h)$

$$\|qI\|_{\widetilde{\Sigma}_h, S}^2 = \inf_{\bar{p} \in P^0(\mathcal{T}_h)} \|(q-\bar{p})I\|_\Omega^2 + \frac{1}{\bar{\lambda}}\|\bar{p}\|_\Omega^2 = \frac{1}{1+\bar{\lambda}}\|qI\|_\Omega^2. \tag{5.44}$$

We can easily prove a lemma on the additive splitting of some $\tau_h \in \widetilde{\Sigma}_h^k$ into a deviatoric and a piecewise constant part, using only algebraic manipulations.

Lemma 5.13. *Let $\tau_h \in \widetilde{\Sigma}_h^k$ be arbitrary, and let $\bar{\tau}_h \in \widetilde{\Sigma}_h^k$ and $p \in P^0(\mathcal{T}_h)$ be such that $\tau_h = \bar{\tau}_h + pI$. Then*

$$\|\tau_h\|_{\widetilde{\Sigma}_h, S}^2 \leq 2\|\bar{\tau}_h\|_\Omega^2 + \sum_{T \in \mathcal{T}_h} h_T^2 \|\operatorname{div} \bar{\tau}_h\|_T^2 + \frac{4}{1+2\bar{\lambda}}\|pI\|_\Omega^2. \tag{5.45}$$

The splitting $\tau_h = \operatorname{dev}_T \tau_h + \frac{1}{d}\operatorname{tr}_T(\tau_h)I$ is orthogonal with respect to the discrete norm,

$$\|\tau_h\|_{\widetilde{\Sigma}_h, S}^2 = \|\operatorname{dev}_T \tau_h\|_{\widetilde{\Sigma}_h, S}^2 + \|\tfrac{1}{d}\operatorname{tr}_T(\tau_h)I\|_{\widetilde{\Sigma}_h, S}. \tag{5.46}$$

5.3.2 Stability

We verify the conditions of Brezzi's theorem, as summarized in Assumption 4.2. First, boundedness of the bilinear forms $a^S(\cdot, \cdot), b(\cdot, \cdot)$ with respect to the discrete, parameter-dependent norms has to be proven. We see that this is straightforward for $a^S(\cdot, \cdot)$, and also holds, but less trivially, for $b(\cdot, \cdot)$.

Lemma 5.14. *Let $\mathcal{T}_h$ be a shape-regular, quasi-uniform triangulation of Ω, and let $a^S(\cdot, \cdot), b(\cdot, \cdot)$ be defined as in (5.37), (5.8). These bilinear forms are bounded on $\widetilde{\Sigma}_h^k$, $\widetilde{\Sigma}_h^k \times (V_h^k \times \Lambda_h^k)$, respectively; the constants of boundedness $\tilde{c}_{a,2}, \tilde{c}_{b,2} > 0$ are independent of $\bar{\lambda} \to \infty$ or $h \to 0$.*

Proof. Continuity of $a^S(\cdot,\cdot)$ follows directly, as

$$a^S(\tau_h,\tau_h) \leq c(d,\bar{\mu})\left[\|\tau_h - \tfrac{1}{d}\operatorname{tr}(\tau_h)I\|_\Omega^2 + \frac{1}{\lambda}\|\tfrac{1}{d}\operatorname{tr}(\tau_h)I\|_\Omega^2 + \sum_{T\in\mathcal{T}_h} h_T^2\|\operatorname{div}\tau_h\|_T^2\right]$$

$$= c(d,\bar{\mu})\left[\inf_{p\in L^2(\mathcal{T}_h)}\|\tau_h - pI\|_\Omega^2 + \frac{1}{\lambda}\|pI\|_\Omega^2 + \sum_{T\in\mathcal{T}_h} h_T^2\|\operatorname{div}\tau_h\|_T^2\right]$$

$$\leq c(d,\bar{\mu})\|\tau_h\|_{\widetilde{\Sigma}_h,S}^2.$$

We proceed to showing continuity for $b(\cdot,\cdot)$. Due to the orthogonality of the splitting $\tau_h = \operatorname{dev}_T\tau_h + \tfrac{1}{d}\operatorname{tr}_T(\tau_h)I$, it is sufficient to prove boundedness of $b(\cdot,\cdot)$ with respect to both parts separately. We first concentrate on the deviatoric part. There we see, using representation (5.9) of $b(\cdot,\cdot)$, the Cauchy-Schwarz inequality and the norm equivalence from Lemma 4.14 applied on $\widetilde{\Sigma}_h^k$

$$b(\operatorname{dev}_T\tau_h;v_h,\mu_h) = \sum_{T\in\mathcal{T}_h}\left[-\int_T \operatorname{dev}_T\tau_h : \varepsilon(v_h)\,dx + \int_{\partial T}(\operatorname{dev}_T\tau_h)_{nn}(v_{h,n}-\mu_{h,n})ds\right]$$

$$\leq \sum_{T\in\mathcal{T}_h}\left[\|\operatorname{dev}_T\tau_h\|_T\|\varepsilon(v_h)\|_T + \|(\operatorname{dev}_T\tau_h)_{nn}\|_{\partial T}\|v_{h,n}-\mu_{h,n}\|_{\partial T}\right]$$

$$\leq c\|\operatorname{dev}_T\tau_h\|_{\widetilde{\Sigma}_h,S}\|v_h,\mu_h\|_{V_h\times\Lambda_h,S}.$$

Next, we consider the trace part $p := \tfrac{1}{d}\operatorname{tr}_T(\tau_h)$. In the estimate below, we use Green's formula (3.12), the trivial scaling inequality $\|q\|_{\partial T} \leq c h_T^{-1/2}\|q\|_T$ for $q \in P^0(\mathcal{T}_h)$, and the Cauchy-Schwarz inequality again. We deduce

$$b(pI;v_h,\mu_h) = -\sum_{T\in\mathcal{T}_h}\int_{\partial T} p\,\mu_h\,dx$$

$$= \inf_{\bar{p}\in P^0(\mathcal{T}_h)}\sum_{T\in\mathcal{T}_h}\left[\underbrace{\int_T \nabla(p-\bar{p})\cdot v\,dx}_{=0} - \int_{\partial T} p\,\mu_{h,n}\,ds\right]$$

$$= \inf_{\bar{p}\in P^0(\mathcal{T}_h)}\sum_{T\in\mathcal{T}_h}\left[-\int_T(p-\bar{p})I:\varepsilon(v_h)dx + \int_{\partial T}(p-\bar{p})(v_{h,n}-\mu_{h,n}) - \bar{p}\mu_{h,n}ds\right]$$

$$\leq c\inf_{\bar{p}\in P^0(\mathcal{T}_h)}\sum_{T\in\mathcal{T}_h}\left[\|(p-\bar{p})I\|_T\|\varepsilon(v_h)\|_T + h_T^{-1/2}\|p-\bar{p}\|_T\|v_{h,n}-\mu_{h,n}\|_{\partial T} + \sqrt{\tfrac{1}{\lambda}}\|\bar{p}I\|_T\sqrt{\tfrac{\bar{\lambda}}{|T|}}\left|\int_{\partial T}\mu_{h,n}\,ds\right|\right]$$

$$\leq c\|pI\|_{\widetilde{\Sigma}_h,S}\|v_h,\mu_h\|_{V_h\times\Lambda_h,S}.$$

From the two estimates above, we can conclude the continuity of $b(\cdot,\cdot)$. □

We now proceed to showing the respective lower bounds, i.e. items 2. (the inf-sup condition) and 3. (coercivity) of Assumption 4.2. We use the following lemma, which bounds the difference between the trace $\operatorname{tr}(\tau_h)$ and its element-wise mean value $\operatorname{tr}_{\mathcal{T}}(\tau_h)$ in terms of deviator and divergence of the finite element function τ_h.

Lemma 5.15. *On a shape-regular, quasi-uniform triangulation $\mathcal{T}_h$ of Ω, let $\widetilde{\Sigma}_h^k$ be the corresponding discontinuous stress finite element space. Then, the difference between $\operatorname{tr}(\tau_h)$ and $\operatorname{tr}_{\mathcal{T}}(\tau_h)$ for $\tau_h \in \widetilde{\Sigma}_h^k$ is bounded element-wise by*

$$\| \operatorname{tr}(\tau_h) - \operatorname{tr}_{\mathcal{T}}(\tau_h) \|_T \leq c \| \operatorname{dev} \tau_h \|_T + h_T \| \operatorname{div} \tau_h \|_T \qquad \text{for } T \in \mathcal{T}_h, \tag{5.47}$$

where $c > 0$ is independent of the local mesh size h_T.

Proof. Let $T \in \mathcal{T}_h$, we note that the difference $\operatorname{tr}(\tau_h) - \operatorname{tr}_{\mathcal{T}}(\tau_h)$ has vanishing mean value on T. Thus, Stokes theory (Lemma 3.39) guarantees the existence of $p_T \in H_0^1(T)$ such that

$$\operatorname{div} p_T = \operatorname{tr}(\tau_h) - \operatorname{tr}_{\mathcal{T}}(\tau_h) \qquad \text{in } T$$

holds in weak sense, and

$$\| \nabla p_T \|_T \leq c \| \operatorname{tr}(\tau_h) - \operatorname{tr}_{\mathcal{T}}(\tau_h) \|_T.$$

We deduce, using that the divergence of the constant function $\operatorname{tr}_{\mathcal{T}}(\tau_h) I$ vanishes, as well as the zero boundary conditions for $p_T \in H_0^1(T)$, when doing integration by parts,

$$\begin{aligned}
\| \operatorname{tr}(\tau_h) - \operatorname{tr}_{\mathcal{T}}(\tau_h) \|_T^2 &= \int_T \operatorname{div} p_T \big(\operatorname{tr}(\tau_h) - \operatorname{tr}_{\mathcal{T}}(\tau_h) \big) \, dx \\
&= - \int_T p_T \cdot \operatorname{div}(\operatorname{tr}(\tau_h) I) \, dx \\
&= -d \int_T p_T \cdot \operatorname{div}(\tau_h - \operatorname{dev} \tau_h) \, dx \\
&= -d \int_T p_T \cdot \operatorname{div} \tau_h \, dx + d \int_T \nabla p_T : \operatorname{dev} \tau_h \, dx.
\end{aligned}$$

Last, we use a scaled Friedrichs inequality for $p_T \in H_0^1(T)$ (Theorem 3.14), and obtain

$$\begin{aligned}
\tfrac{1}{d} \| \operatorname{tr}(\tau_h) - \operatorname{tr}_{\mathcal{T}}(\tau_h) \|_T^2 &\leq \| p_T \|_T \| \operatorname{div} \tau_h \|_T + \| \nabla p_T \|_T \| \operatorname{dev} \tau_h \|_T \\
&\leq (1 + c_F) \| \nabla p_T \|_T \big(\| \operatorname{dev} \tau_h \|_T + h_T \| \operatorname{div} \tau_h \|_T \big).
\end{aligned}$$

□

Lemma 5.16. *Let $\mathcal{T}_h$ be a shape-regular, quasi-uniform triangulation of Ω. Then, $a^S(\cdot,\cdot)$ as defined in (5.37) is coercive on $\widetilde{\Sigma}_h^k$ with respect to $\| \cdot \|_{\widetilde{\Sigma}_h, S}$, there exists a constant $\tilde{c}_{a,1} > 0$ independent of the mesh size h or the Lamé parameter $\bar{\lambda} \to \infty$ such that*

$$a^S(\tau_h, \tau_h) \geq \tilde{c}_{a,1} \| \tau_h \|_{\widetilde{\Sigma}_h, S}^2 \qquad \forall \tau_h \in \widetilde{\Sigma}_h^k. \tag{5.48}$$

Proof. Coercivity of $a^S(\cdot,\cdot)$ can easily be shown using Lemma 5.15. To estimate $\|\tau_h\|_{\tilde{\Sigma}_h,S}$ for some $\tau_h \in \tilde{\Sigma}_h^k$, we bound the infimum in the definition of the norm by setting $\bar{p} = \frac{1}{d}\operatorname{tr}_T(\tau_h)$. We obtain

$$\|\tau_h\|_{\tilde{\Sigma}_h,S}^2 \leq \|\tau_h - \tfrac{1}{d}\operatorname{tr}_T(\tau_h)I\|_\Omega^2 + \frac{1}{d\bar{\lambda}}\|\operatorname{tr}_T(\tau_h)\|_\Omega^2 + \sum_{T\in\mathcal{T}_h} h_T^2 \|\operatorname{div}\tau_h\|_T^2.$$

The first term can be estimated, inserting $\operatorname{tr}(\tau_h)$ and using estimate (5.47) in Lemma 5.15

$$\tfrac{1}{2}\|\tau_h - \tfrac{1}{d}\operatorname{tr}_T(\tau_h)I\|_\Omega^2 \leq \|\underbrace{\tau_h - \tfrac{1}{d}\operatorname{tr}(\tau_h)I}_{=\operatorname{dev}\tau_h}\|_\Omega^2 + \tfrac{1}{d}\|\operatorname{tr}(\tau_h) - \operatorname{tr}_T(\tau_h)\|_\Omega^2$$

$$\leq c\Big(\|\operatorname{dev}\tau_h\|_\Omega^2 + \sum_{T\in\mathcal{T}_h} h_T^2 \|\operatorname{div}\tau_h\|_T^2\Big).$$

Inserting this into the first estimate yields

$$\|\tau_h\|_{\tilde{\Sigma}_h,S}^2 \leq c\Big(\|\operatorname{dev}\tau_h\|_\Omega^2 + \sum_{T\in\mathcal{T}_h} h_T^2 \|\operatorname{div}\tau_h\|_T^2 + \frac{1}{\bar{\lambda}}\|\operatorname{tr}_T(\tau_h)\|_\Omega^2\Big)$$

$$\leq c\, a^S(\tau_h,\tau_h).$$

Here we used that $\|\operatorname{tr}_T(\tau_h)\|_\Omega \leq c\|\operatorname{tr}(\tau_h)\|_\Omega$. The constant c does not depend on $h \to 0$ or $\bar{\lambda} \to \infty$. $\square$

Lemma 5.17. *Let $\mathcal{T}_h$ be a shape-regular, quasi-uniform triangulation of Ω. Then, $b(\cdot,\cdot)$ as defined as in (5.8) satisfies an inf-sup condition with respect to the parameter-dependent norms $\|\cdot\|_{\tilde{\Sigma}_h,S}, \|\cdot\|_{V_h\times\Lambda_h,S}$. The constant $\tilde{c}_{b,1} > 0$ is independent of the mesh size h or the Lamé parameter $\bar{\lambda} \to \infty$, such that*

$$\inf_{\tau_h\in\tilde{\Sigma}_h^k} \sup_{\substack{v_h\in V_h^k \\ \mu_h\in\Lambda_h^k}} \frac{b(\tau_h;v_h,\mu_h)}{\|\tau_h\|_{\tilde{\Sigma}_h,S}\|v_h,\mu_h\|_{V_h\times\Lambda_h,S}} \geq \tilde{c}_{b,1}. \tag{5.49}$$

Proof. The proof of the inf-sup condition in the almost incompressible setting relies on the one in case of compressible materials, Lemma 4.17. For given $(v_h,\mu_h) \in V_h^k \times \Lambda_h^k$, one can find $\bar{\tau}_h \in \tilde{\Sigma}_h^k$ such that

$$b(\bar{\tau}_h;v_h,\mu_h) \geq c\|\bar{\tau}_h\|_\Omega \Big(\sum_{T\in\mathcal{T}_h}\big[\|\varepsilon(v_h)\|_T^2 + h_T^{-1}\|v_{h,n} - \mu_{h,n}\|_{\partial T}^2\big]\Big)^{1/2},$$

and

$$\|\bar{\tau}_h\|_\Omega^2 \leq c\sum_{T\in\mathcal{T}_h}\big[\|\varepsilon(v_h)\|_T^2 + h_T^{-1}\|v_{h,n} - \mu_{h,n}\|_{\partial T}^2\big]. \tag{5.50}$$

Now, we define the piecewise constant function $p \in P^0(\mathcal{T}_h)$ element-wise via

$$p|_T := -\frac{\bar{\lambda}}{|T|}\int_{\partial T} \mu_{h,n}\, ds.$$

This choice implies

$$b(pI; v_h, \mu_h) = -\sum_{T \in \mathcal{T}_h} \int_{\partial T} p\mu_{h,n} \, ds = \sum_{T \in \mathcal{T}_h} \frac{\bar{\lambda}}{|T|} \left| \int_{\partial T} \mu_{h,n} \, ds \right|^2. \tag{5.51}$$

We choose $\tau_h := \bar{\tau}_h + pI$, and see

$$b(\tau_h; v_h, \mu_h) = b(\bar{\tau}_h; v_h, \mu_h) + b(pI; v_h, \mu_h) \geq c \, \|v_h, \mu_h\|^2_{V_h \times \Lambda_h, S}.$$

Therefore, bounding $\|\tau_h\|_{\bar{\Sigma}_h, S}$ by $\|v_h, \mu_h\|_{V_h \times \Lambda_h, S}$ from above concludes the proof. We use estimate (5.45) in Lemma 5.13 and the bound (5.50) together with an inverse inequality for div $\bar{\tau}_h$, and obtain

$$\|\bar{\tau}_h + pI\|_{\bar{\Sigma}_h, S} \overset{(5.45)}{\leq} 2\|\bar{\tau}_h\|^2_\Omega + \sum_{T \in \mathcal{T}_h} h_T^2 \|\operatorname{div} \bar{\tau}_h\|^2_T + \frac{4}{1+2\bar{\lambda}} \|pI\|^2_\Omega$$

$$= 2\|\bar{\tau}_h\|^2_\Omega + \sum_{T \in \mathcal{T}_h} \left[h_T^2 \|\operatorname{div} \bar{\tau}_h\|^2_T + \frac{4d\bar{\lambda}^2}{|T|(1+2\bar{\lambda})} \left| \int_{\partial T} \mu_{h,n} \, ds \right|^2 \right]$$

$$\overset{(5.50)}{\leq} c \sum_{T \in \mathcal{T}_h} \left[\|\varepsilon(v_h)\|^2_T + h_T^{-1} \|v_{h,n} - \mu_{h,n}\|^2_{\partial T} + \frac{\bar{\lambda}}{|T|} \left| \int_{\partial T} \mu_{h,n} \, ds \right|^2 \right]$$

$$= \|v_h, \mu_h\|_{V_h \times \Lambda_h, S}.$$

This implies the stability estimate (5.49). $\qquad\square$

Summarizing our results, we see that Lemmas 5.14, 5.16 and 5.17 ensure that all conditions for the application of Brezzi's theory (i.e. of Assumption 4.2) are satisfied. This implies an a-priori bound for the discretization error in the discrete norms. In the following, we use this bound to estimate the error by quantities, which are independent of $\bar{\lambda}$, and contain only derivatives of the solution (σ, u).

Theorem 5.18. *Let (σ, u) be a solution to the equations of elasticity, as posed in Problem 3.1, and let λ denote the normal displacement on the set of interfaces $\mathcal{F}_h$, i.e. $\lambda|_F := u_{n_F}$. Let $(\sigma_h, u_h, \lambda_h)$ be a solution to the hybridized set of discrete equations in Problem 5.1. Let $m \leq k$ be a positive integer, such that $\sigma \in H^m_{SYM}(\Omega)$, and $u \in [H^{m+1}(\Omega)]^d$. Then*

$$\|\sigma - \sigma_h\|_{\bar{\Sigma}_h, S} + \|u - u_h, \lambda - \lambda_h\|_{V_h \times \Lambda_h, S} \leq c h^m \Big(\|\nabla^m \sigma\|_\Omega + \|\nabla^m \varepsilon(u)\|_\Omega \Big), \tag{5.52}$$

where the constant c is independent of the mesh size h, and does not deteriorate for almost incompressible materials, i.e. for $\bar{\lambda} \to \infty$.

Proof. Due to the basic error estimate for mixed problems provided in Lemma 4.3, we may estimate

$$\|\sigma - \sigma_h\|^2_{\widetilde{\Sigma}_h,S} + \|u - u_h, \lambda - \lambda_h\|^2_{V_h \times \Lambda_h, S} \leq c \left(\inf_{\tau_h \in \widetilde{\Sigma}_h^k} \|\sigma - \tau_h\|^2_{\widetilde{\Sigma}_h,S} + \right. \tag{5.53}$$

$$\left. \inf_{\substack{v_h \in V_h^k \\ \mu_h \in \Lambda_h^k}} \sum_{T \in \mathcal{T}_h} \left[\|\varepsilon(u - v_h)\|^2_T + h_T^{-1}\|v_{h,n} - \mu_{h,n}\|^2_{\partial T} + \frac{\bar{\lambda}}{|T|} \left| \int_{\partial T} (u_n - \mu_{h,n}) ds \right|^2 \right] \right). \tag{5.54}$$

The first term containing the best-approximation error for the stress space can be estimated by standard arguments, as $\widetilde{\Sigma}_h^k$ is piecewise polynomial and discontinuous. For bounding the second infimum, i.e. line (5.54), we employ special choices of v_h, μ_h. We set v_h to the nodal Nédélec interpolant of u, i.e. $v_h := \mathcal{I}_{h,k}^V u$. For the Lagrangian multiplier, we choose $\mu_h := \Pi^k \lambda = \Pi^k u_n$ as the facet-wise L^2 projection onto $P^k(\mathcal{F}_h)$. We obtain

$$(5.54) \leq \sum_{T \in \mathcal{T}_h} \left[\|\varepsilon(u - \mathcal{I}_{h,k}^V u)\|^2_T + 2h_T^{-1}(\|u_n - \mathcal{I}_{h,k}^V u_n\|^2_{\partial T} + \|u_n - \Pi^k u_n\|^2_{\partial T}) \right.$$

$$\left. + \frac{\bar{\lambda}}{|T|} \underbrace{\left| \int_{\partial T} (u_n - \Pi^k u_n) ds \right|^2}_{=0} \right].$$

Here, we used that the mean value of $u_n - \Pi^k u_n$ vanishes on each facet of the triangulation. Now, employing the approximation properties of the interpolation operator $\mathcal{I}_{h,k}^V$ (Theorem 4.21) and the L^2 projection Π^k, we obtain

$$(5.54) \leq c h^{2m} \|\nabla^m \varepsilon(u)\|^2_\Omega.$$

This estimate then completes the proof. □

Similar to the compressible case, the stability estimates ensure the existence of a Schur complement operator $S_h^S = B_h(A_h^S)^{-1} B_h^*$, where A_h^S is the discrete linear operator corresponding to the stabilized bilinear form $a^S(\cdot, \cdot)$. Then, $s^S(\cdot, \cdot)$ shall denote the corresponding bilinear form on the discrete space $V_h^k \times \Lambda_h^k$. For the right hand side $H_h^S := B_h(A_h^S)^{-1} F_1^S - F_2$, the stabilized hybrid system (5.39) is equivalent to finding

$$s^S(u_h, \lambda_h; v_h, \mu_h) = \langle H_h^S; v_h, \mu_h \rangle_{V_h^k \times \Lambda_h^k} \quad \forall v_h \in V_h^k, \mu_h \in \Lambda_h^k, \tag{5.55}$$

and setting $\sigma_h \in \widetilde{\Sigma}_h^k$ such that

$$a^S(\sigma_h, \tau_h) = \langle F_1^S; \tau_h \rangle_\Sigma - b(\tau_h; u_h, \lambda_h) \quad \forall \tau_h \in \widetilde{\Sigma}_h^k.$$

The Schur complement S_h^S is symmetric and positive definite on $V_h^k \times \Lambda_h^k$, and spectrally equivalent to the stabilized norm,

$$s^S(v_h, \mu_h; v_h, \mu_h) \simeq \|v_h, \mu_h\|^2_{V_h \times \Lambda_h, S}. \tag{5.56}$$

One can therefore solve system (5.39) by a preconditioned CG method. We propose to use the additive Schwarz block preconditioner constructed in Section 5.2. Subsequently, we will show that it is also suitable in case of nearly incompressible materials.

5.3.3 A preconditioner for nearly incompressible materials

In Section 5.2, we proposed an additive Schwarz block preconditioner to use in a preconditioned CG method in order to solve Problem 5.1. We will now show that this choice also works fine in case of nearly incompressible materials, provided the basis for the Lagrangian multiplier space Λ_h^k is orthogonal with respect to the L^2 inner product. To this end, we assume that $\mathcal{T}_h$ is a shape-regular triangulation of Ω. For $v_h \in V_h^k, \mu_h \in \Lambda_h^k$, let

$$v_h = v_0 + \sum_{X \in \mathcal{X}_h} v_X, \qquad \mu_h = \mu_0 + \sum_{F \in \mathcal{F}_h} \mu_F$$

be the corresponding decompositions, as given in (5.30), (5.31). Let E_0, E_X^V, E_F^Λ be the respective restriction operators, as defined in equation (5.32). We denote the sub-blocks of the stabilized Schur complement S_h^S from equation (5.55) by

$$S_0^S := E_0^* S_h^S E_0, \qquad S_X^{S,V} := \left(E_X^V\right)^* S_h^S E_X^V, \qquad S_F^{S,\Lambda} := \left(E_F^\Lambda\right)^* S_h^S E_F^\Lambda.$$

This allows to introduce the preconditioner

$$\left(C_h^S\right)^{-1}(v_h, \mu_h) := \left(S_0^S\right)^{-1}(v_{h,0}, \mu_{h,0}) + \sum_{X \in \mathcal{X}_h} \left(S_X^{S,V}\right)^{-1}(v_X, 0) + \sum_{F \in \mathcal{F}_h} \left(S_F^{S,\Lambda}\right)^{-1}(0, \mu_F). \tag{5.57}$$

The stabilized Schur complement is spectrally equivalent to C_h^S, as the next theorem states.

Theorem 5.19. *On a shape-regular, quasi-uniform triangulation $\mathcal{T}_h$ of Ω, let S_h^S denote the Schur complement operator from equation (5.55), and let $(C_h^S)^{-1}$ be the corresponding additive Schwarz block preconditioner defined by equation (5.57). Then S_h^S and C_h^S are spectrally equivalent, provided the finite element basis for Λ_h^k is orthogonal with respect to the L^2 inner product,*

$$\langle S_h^S(v_h, \mu_h); v_h, \mu_h \rangle \simeq \langle C_h^S(v_h, \mu_h); v_h, \mu_h \rangle \qquad \forall v_h \in V_h^k, \mu_h \in \Lambda_h^k. \tag{5.58}$$

The constants of equivalence do not depend on the mesh size h.

Proof. The proof is similar to the proof of Theorem 5.8, which stated the same relation between S_h and C_h in the compressible setting. There, we argued, that it is sufficient to show

$$\|v_h, \mu_h\|_{V_h \times \Lambda_h, S} \simeq \|v_0, \mu_0\|_{V_h \times \Lambda_h, S} + \sum_{X \in \mathcal{X}_h} \|v_X, 0\|_{V_h \times \Lambda_h, S} + \sum_{F \in \mathcal{F}_h} \|0, \mu_F\|_{V_h \times \Lambda_h, S}. \tag{5.59}$$

We observe that the squared norms $\|\cdot\|^2_{V_h\times\Lambda_h}$ and $\|\cdot\|^2_{V_h\times\Lambda_h,S}$ differ by the additive term

$$\sum_{T\in\mathcal{T}_h}\left|\int_{\partial T}\mu_{h,n}\,ds\right|^2.$$

When using an L^2 orthogonal basis for Λ^k_h, the mean value of high-order components of μ_h on a facet $F\in\mathcal{F}_h$ vanishes, and thus

$$\left|\int_F \mu_{h,n}\,ds\right| = \left|\int_F \mu_{0,n}\,ds\right|.$$

We first bound the different contributions on the right hand side of (5.59) by $\|v_h,\mu_h\|^2_{V_h\times\Lambda_h,S}$. Using Lemma 5.9, we may estimate the low-order contribution

$$\begin{aligned}\|v_0,\mu_0\|^2_{V_h\times\Lambda_h,S} &= \|v_0,\mu_0\|_{V_h\times\Lambda_h} + \sum_{T\in\mathcal{T}_h}\left|\int_{\partial T}\mu_{0,n}\,ds\right|^2\\ &\leq c\|v_h,\mu_h\|_{V_h\times\Lambda_h} + \sum_{T\in\mathcal{T}_h}\left|\int_{\partial T}\mu_{h,n}\,ds\right|^2 = c\,\|v_h,\mu_h\|^2_{V_h\times\Lambda_h,S}.\end{aligned}$$

Due to the fact that μ_F is orthogonal to the piecewise constant functions, $\int_F \mu_{F,n}\,ds$ vanishes. Therefore, for the high-order parts $(v_X,0)$ and $(0,\mu_F)$, the original and stabilized norm coincide. Thus, Lemma 5.10 ensures

$$\sum_{X\in\mathcal{X}_h}\|v_X,0\|_{V_h\times\Lambda_h,S} + \sum_{F\in\mathcal{F}_h}\|0,\mu_F\|_{V_h\times\Lambda_h,S} \leq c\|v_h,\mu_h\|_{V_h\times\Lambda_h} \leq c\|v_h,\mu_h\|_{V_h\times\Lambda_h,S}.$$

From these observations we conclude, that the sum of all block contributions is bounded from above by $\|v_h,\mu_h\|_{V_h\times\Lambda_h,S}$. For the lower bound, we employ, similar to Lemma 5.11, the argument of finite overlap of support for the blocks of v_h, and disjoint supports of the facet-blocks of μ_h. Collecting our results, we arrive at the statement of the theorem. □

5.4 Numerical results

We present some results obtained using the hybridized TD-NNS method. Again, all computations were done using the finite element package Netgen/NgSolve, [Sch97]. As a first example, we consider the unit square, fixed on the left hand side ($u = 0$), and where surface tractions $\sigma_n = (1,1)^T$ are applied on the right hand side. We assume that the material is close to the incompressible limit, and set Young's modulus $\bar{E} = 1$, and Poisson's ratio $\bar{\nu} = 0.499\,999$. We use a triangulation consisting of 114 elements. We compute the error $\|\sigma-\sigma_h\|_\Omega$ for polynomial orders $k = 1, 2, 4$, doing adaptive mesh refinement using a Zienkiewicz-Zhu type error estimator [ZZ87]. In Figure 5.1, we plot the obtained results. We see an optimal rate of convergence of the error $\|\sigma-\sigma_h\|_\Omega$. Note that

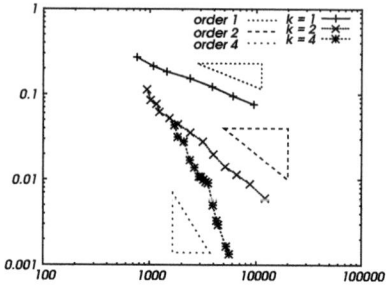

Figure 5.1: Coupling degrees of freedom vs. error $\|\sigma - \sigma_h\|_\Omega$, almost incompressible material (Young's modulus $\bar{E} = 1$, Poisson ratio $\bar{\nu} = 0.499\,999$): We observe an optimal rate of convergence for methods of orders $k = 1, 2, 4$ and adaptive refinement.

	orthogonal		non-orthogonal	
ϵ	$k=2$	$k=5$	$k=2$	$k=5$
10^{-1}	38	60	61	169
10^{-2}	43	77	88	239
10^{-3}	45	82	164	294
10^{-4}	45	83	240	796
10^{-7}	45	83	398	2777
10^{-10}	45	83	–	–

Table 5.1: Multiplicative Schwarz block preconditioner, number of iterations in PCG method for almost incompressible material (Young's modulus $\bar{E} = 1$, Poisson ratio $\bar{\nu} = 0.5 - \epsilon$) on the unit square, using orthogonal and non-orthogonal basis functions for Λ_h^k

in the analysis, we were only able to provide the corresponding bound for $\|\sigma - \sigma_h\|_{\widetilde{\Sigma}_h, S}$.

Next, we investigate on the number of iterations needed in the preconditioned CG iteration, where we use a multiplicative version of the additive Schwarz block preconditioner provided in Section 5.2. In our computations, we keep the mesh fixed to 114 elements, and choose polynomial orders $k = 2, 5$. This results in 1 134 versus 2 268 coupling degrees of freedom. We set Poisson's ratio to $\bar{\nu} = 0.5 - \epsilon$, where ϵ ranges between 0.1 and 10^{-10}. In Section 5.3.3, we required the basis for Λ_h^k to be orthogonal with respect to the L^2 inner product. We now do computations using orthogonal and non-orthogonal bases. We see that the number of iterations is only independent of ϵ, if the basis is chosen orthogonally. If not, the PCG solver needs more steps as the Poisson ratio tends to $1/2$. In Table 5.1, we list the numbers of iterations needed to reduce the error by a factor of 10^{-12}. The PCG solver does not converge within 3 000 iterations, when using non-orthogonal basis functions.

In Figure 5.2, we plot the dependence of the block preconditioner on the mesh size h. We start

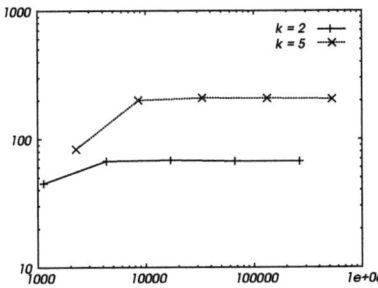
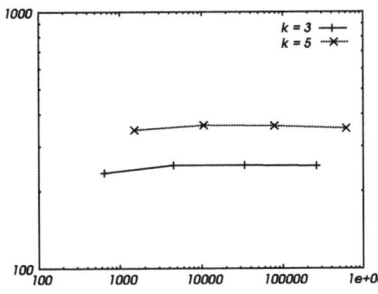

Figure 5.2: Coupling degrees of freedom vs. number of iterations, almost incompressible material, (Young's modulus $\bar{E} = 1$, Poisson ratio $\bar{\nu} = 0.499\,999$), uniform mesh refinement; left two space dimensions, right three space dimensions.

	orthogonal		non-orthogonal	
ϵ	$k=3$	$k=5$	$k=3$	$k=5$
10^{-1}	123	187	200	271
10^{-2}	141	221	296	413
10^{-3}	148	236	521	784
10^{-4}	149	235	905	1497
10^{-7}	149	236	2971	–
10^{-10}	159	234	–	–

Table 5.2: Multiplicative Schwarz block preconditioner, number of iterations in PCG method for almost incompressible material (Young's modulus $\bar{E} = 1$, Poisson ratio $\bar{\nu} = 0.5 - \epsilon$) on the unit cube, using orthogonal and non-orthogonal basis functions for Λ_h^k

from a grid consisting of 114 triangles, and do uniform refinement. We use methods of orders $k = 2, 5$, and obtain numbers of iterations which are independent of the decreasing mesh size.

We provide results for a similar example in three space dimensions: We consider the unit cube, one side fixed, constant surface tractions acting on the opposite side. We assume to be given a Young's modulus $\bar{E} = 1$, and varying Poisson ratio $\bar{\nu} = 0.5 - \epsilon$. We study the behavior of the preconditioner with respect to mesh refinement and near incompressibility. We use a coarse mesh consisting of 12 tetrahedral elements. The PCG method is terminated, when an error reduction of 10^{-12} is achieved. In Table 5.2, numbers of iterations for methods of orders $k = 3, 5$ and ϵ varying between 0.1 and 10^{-10} are listed. We use both orthogonal and non-orthogonal basis functions for the facet space. For a non-orthogonal basis, the method does not converge within 3 000 iterations for ϵ sufficiently small. In Figure 5.2 on the right hand side, we plot the number of iterations needed after doing uniform mesh refinement.

Chapter 6

Anisotropic elements

This chapter is devoted to the application of the TD-NNS method on anisotropic domains. Examples for such domains are thin plates or shells. Discretizing thin structures with shape-regular elements leads to an enormous amount of degrees of freedom, as then the mesh size has to be of the order of the thickness of the domain. It seems more natural to use meshes consisting of flat tensor product elements, such as quadrilaterals, prisms or hexahedrons. However, many standard methods, such as the pure displacement formulation, break down in such a setting. One can trace these effects back to the constant of stability, which directly depends on the constant in Korn's inequality. It deteriorates, when the aspect ratio of the elements becomes large. In the engineering literature, this effect is referred to as "shear locking".

A possible remedy to this problem are plate and shell models. In these methods, the full, three-dimensional problem is reduced to a 2D model on the mid-surface of the domain. Well known models for plates are the Reissner-Mindlin [Rei45, Min51] or the Kirchhhoff plate models [TW59]. For shells, a widely used model for discretization is the Koiter model [Koi60].

A generalization of such plate or shell models is the idea of hierarchical modeling. There, a polynomial dependence of the quantities of interest on the thickness direction is assumed. Thereby, one obtains a family of models, as one increases the polynomial order. This may be seen as a p-version of a related, three-dimensional finite element method. Such models were introduced in variational form by [VB81]. An overview over the different discretization techniques within a unified framework is provided in [DFY04]. Analysis for the approach by hierarchical models can be found in [AAFM99, BL91, DBR01], a-posteriori error estimates in [Ain98, Sch96, SO97].

In our approach, we do not aim at reducing the dimension of the problem. We rather want to use anisotropic, flat, quadrilateral or prismatic finite elements for slim domains in two- or three-dimensional space. This way, we can save a considerable amount of degrees of freedom when compared to standard methods, and have more freedom in choosing the order of approximation with respect to in-plane and thickness discretizations element-wise.

In the following, we assume our body to be of tensor product structure, such as a beam, plate or disc. We use a triangulation consisting of quadrilateral or prismatic elements, which is aligned with the structure of the body. We show that the constant of stability for the TD-NNS method is independent of the aspect ratio of the domain or the finite elements. Moreover, we provide interpolation operators, which then imply an optimal order of convergence of the discrete solution. We emphasize that our estimates do not deteriorate, as the aspect ratio of the domain or the finite elements grows.

Section 6.1 deals with setup and application of the TD-NNS method on an anisotropic domain. Finite elements for prisms and quadrilaterals are provided, stability of the method independently of the anisotropic mesh sizes h_x, h_z and their ratio h_x/h_z is shown. In Section 6.2, a family of quasi-interpolation operators for tensor product spaces is introduced. Their approximation properties are investigated. Finally, numerical results are provided in Section 6.3. These calculations imply the applicability of the method also on curved domains.

6.1 The TD-NNS method on an anisotropic domain

In this section, we are concerned with the setup of the TD-NNS method on a geometrically anisotropic domain and its stability properties. We consider domains of tensor product structure, such as thin plates or discs. We adapt our notation to this setting, where in-plane and transversal components are of different quality. We use a tensor product mesh of quadrilateral or prismatic elements, which is a natural choice given the structure of the domain. We derive stability estimates and provide finite elements for these element types.

6.1.1 Anisotropic setting

We first specify the notion of a tensor product domain.

Definition 6.1. *We call $\Omega \subset \mathbb{R}^d, d = 2, 3$ of tensor product structure, if it is of the form $\Omega = \Omega_x \times \Omega_z$. There $\Omega_x \subset \mathbb{R}^{d-1}$ corresponds to the cross section. It is a bounded, connected domain of size $d_x := \text{diam}(\Omega_x)$. The second component, $\Omega_z \subset \mathbb{R}$, then describes the thickness direction, and is an interval of length d_z. In three dimensions, we assume that Ω_x is a polygonal Lipschitz domain satisfying Assumption 3.2.*

Let $\Gamma_x = \partial \Omega_x$, $\Gamma_z = \partial \Omega_z$ be the respective boundaries of cross section and thickness. For simplicity of notation, we restrict ourselves to the Dirichlet problem throughout this chapter. Then all finite element spaces can be designed as tensor products of spaces on the line or in the plane. The case of mixed boundary conditions can be deduced easily, but one needs to impose the different types of conditions additionally.

In many applications, d_z is much smaller than d_x. To emphasize the different qualities of the directions, we use coordinates $(\boldsymbol{x}, z) = (x_1, z)$ in two or $(\boldsymbol{x}, z) = (x_1, x_2, z)$ in three space dimensions. For a vector-valued function v, we refer to its components by

$$\begin{pmatrix} v_x \\ v_z \end{pmatrix} = \begin{cases} (v_{x_1}, v_z)^T & \text{if } d = 2, \\ (v_{x_1}, v_{x_2}, v_z)^T & \text{if } d = 3. \end{cases}$$

A tensor-valued symmetric function τ can be divided into four sub-blocks

$$\tau = \begin{pmatrix} \tau_{\boldsymbol{x}} & \tau_{\boldsymbol{x}z} \\ \tau_{\boldsymbol{x}z}^T & \tau_z \end{pmatrix}.$$

The lower right block τ_z is always scalar-valued, whereas $\tau_{\boldsymbol{x}}, \tau_{\boldsymbol{x}z}$ are of the form

$$\tau_{\boldsymbol{x}} = \begin{cases} \tau_{x_1 x_1} & \text{if } d = 2, \\ \begin{pmatrix} \tau_{x_1 x_1} & \tau_{x_1 x_2} \\ \tau_{x_1 x_2} & \tau_{x_2 x_2} \end{pmatrix} & \text{if } d = 3, \end{cases} \qquad \tau_{\boldsymbol{x}z} = \begin{cases} \tau_{x_1 z} & \text{if } d = 2, \\ (\tau_{x_1 z}, \tau_{x_2 z})^T & \text{if } d = 3. \end{cases}$$

Similarly, $\varepsilon_{\boldsymbol{x}}(v)$, $\varepsilon_z(v)$ and $\varepsilon_{\boldsymbol{x}z}(v)$ denote the respective sub-blocks of the strain tensor $\varepsilon(v)$ of a vector-valued function v.

6.1.1.1 Anisotropic triangulation

In order to exploit the tensor product nature of the domain Ω, we introduce a matching triangulation. Let therefore

$$T_{h_{\boldsymbol{x}}}^{\boldsymbol{x}} = \{T^{\boldsymbol{x}}\}, \qquad T_{h_z}^z = \{T^z\}$$

be simplicial, uniform and shape regular triangulations of $\Omega_{\boldsymbol{x}}, \Omega_z$, respectively. Thus, $T_{h_{\boldsymbol{x}}}^{\boldsymbol{x}}$ is a triangular mesh in three dimensions, or a subdivision in two dimensions. In both cases $T_{h_z}^z$ is a subdivision of the line segment Ω_z. The parameters $h_{\boldsymbol{x}}, h_z$ denote the respective mesh sizes. For both triangulations, we define the set of element interfaces

$$\mathcal{F}_{h_{\boldsymbol{x}}}^{\boldsymbol{x}} = \{F^{\boldsymbol{x}}\}, \qquad \mathcal{F}_{h_z}^z = \{F^z\}.$$

In two space dimensions, both sets correspond to the sets of points defining the respective subdivisions, in three dimensions $\mathcal{F}_{h_{\boldsymbol{x}}}^{\boldsymbol{x}}$ consists of triangle edges in the plane.

Using these quantities, we define a tensor product mesh

$$\mathcal{T}_h = \{T = T^{\boldsymbol{x}} \times T^z : T^{\boldsymbol{x}} \in T_{h_{\boldsymbol{x}}}^{\boldsymbol{x}}, T^z \in T_{h_z}^z\}.$$

This is now a quadrilateral mesh in two, or a prismatic mesh in three space dimensions by definition. Let $\mathcal{F}_h$ be the set of element interfaces for the triangulation $\mathcal{T}_h$. This set can be split into an in-plane part $\mathcal{F}_\parallel$ and a vertical part $\mathcal{F}_\perp$,

$$\begin{aligned} \mathcal{F}_h &:= \mathcal{F}_\parallel \cup \mathcal{F}_\perp, \\ \mathcal{F}_\parallel &:= \{T^{\boldsymbol{x}} \times F^z : T^{\boldsymbol{x}} \in T_{h_{\boldsymbol{x}}}^{\boldsymbol{x}}, F^z \in \mathcal{F}_{h_z}^z\}, \\ \mathcal{F}_\perp &:= \{F^{\boldsymbol{x}} \times T^z : F^{\boldsymbol{x}} \in \mathcal{F}_{h_{\boldsymbol{x}}}^{\boldsymbol{x}}, T^z \in T_{h_z}^z\}. \end{aligned}$$

Thus, the in-plane part $\mathcal{F}_\parallel$ consists of all horizontal edges in two dimensions, and of in-plane, triangular facets in three dimensions. The subset $\mathcal{F}_\perp$ contains all vertical facets, which are edges in two and quadrilateral faces in three dimensions.

In Chapter 4, we defined the local mesh size for a facet F by

$$h_F := \bigl(|T_1| + |T_2|\bigr)/|F|,$$

where T_1, T_2 were the elements adjacent to F. We note, that in the anisotropic setting, where both triangulations $\mathcal{T}_{h_x}^x$ and $\mathcal{T}_{h_z}^z$ are assumed to be uniform and shape regular, this parameter is either of size h_x or h_z. There holds

$$h_F \simeq \begin{cases} h_x & \text{if } F \in \mathcal{F}_\perp, \\ h_z & \text{if } F \in \mathcal{F}_\parallel. \end{cases}$$

6.1.2 Finite element spaces

In the following, we propose finite element spaces V_h^k, Σ_h^k based on the tensor product mesh $\mathcal{T}_h$. For the construction, we use the finite element spaces for H^1, $H(\mathrm{curl})$, $H(\mathrm{div\,div})$ and L^2 from Chapter 4 on the simplicial triangulations $\mathcal{T}_{h_x}^x$ and $\mathcal{T}_{h_z}^z$. The way the finite element spaces are defined indicates how to choose the corresponding shape functions. This will then be done in Section 6.1.3.

6.1.2.1 In-plane and transversal finite element spaces

In Chapter 4, we introduced finite element spaces with different continuity conditions on a triangular mesh. We now define these spaces for the triangulation $\mathcal{T}_{h_x}^x$ of the cross section Ω_x. We use x as an index, to indicate that the spaces correspond to Ω_x and $\mathcal{T}_{h_x}^x$. In case of three space dimensions, we define

$$W_x^k := \{w_h \in \mathcal{C}(\Omega_x) : w_h \in P^k(\mathcal{T}_{h_x}^x),\ w_h = 0 \text{ on } \Gamma_x\}, \tag{6.1}$$

$$V_x^k := \{v_h \in [L^2(\Omega_x)]^2 : v_h \in [P^k(\mathcal{T}_{h_x}^x)]^d,\ v_{h,\tau} \text{ cont.},\ v_{h,\tau} = 0 \text{ on } \Gamma_x\}, \tag{6.2}$$

$$\Sigma_x^k := \{\tau_h \in L^2_{\mathrm{SYM}}(\Omega_x) : \tau_h \in P^k_{\mathrm{SYM}}(\mathcal{T}_{h_x}^x),\ \tau_{h,nn} \text{ cont.}\}, \tag{6.3}$$

$$\mathcal{P}_x^k := \{q_h \in L^2(\Omega_x) : q_h \in P^k(\mathcal{T}_{h_x}^x)\}. \tag{6.4}$$

On the line segment Ω_z, we also define the respective spaces, now indexed with z. Due to the one-dimensionality of Ω_z, the different continuity conditions reduce to continuous and non-continuous spaces,

$$W_z^k := \{w_h \in \mathcal{C}(\Omega_z) : w_h \in P^k(\mathcal{T}_{h_z}^z),\ w_h = 0 \text{ on } \Gamma_z\}, \tag{6.5}$$

$$\Sigma_z^k := \{\tau_h \in \mathcal{C}(\Omega_z) : \tau_h \in P^k(\mathcal{T}_{h_z}^z)\}, \tag{6.6}$$

$$\mathcal{P}_z^k := V_z^k := \{v_h \in L^2(\Omega_z) : v_h \in P^k(\mathcal{T}_{h_z}^z)\}. \tag{6.7}$$

For $d = 2$, Ω^x is a line segment. In this case, the finite element spaces $W_x^k, \Sigma_x^k, V_x^k = \mathcal{P}_x^k$ are defined as the spaces of piecewise polynomial, continuous and non-continuous functions, as done above for W_z^k, Σ_z^k, and $V_z^k = \mathcal{P}_z^k$. Degrees of freedom and shape functions for the different spaces in two space dimensions were provided in Chapter 4, the definition of similar finite elements on the line segment are straightforward and not given in detail in this thesis.

6.1.2.2 Tensor product spaces

The global finite element spaces are now defined using the in-plane and transversal spaces from above. For the displacements, we use the standard Nédélec space on a tensor product mesh, as one can find e.g. in [Mon03],

$$V_h^k := \{v_h \in [L^2(\Omega)]^3 \,:\, v_{h,\boldsymbol{x}} \in V_{\boldsymbol{x}}^k \otimes W_z^{k+1}, v_{h,z} \in W_{\boldsymbol{x}}^{k+1} \otimes V_z^k\}. \tag{6.8}$$

For the stresses, the definition of Σ_h^k differs slightly for two and three space dimensions. This is implied by the different nature of shape functions for the two- and one-dimensional spaces, as we will see in the proof of Lemma 6.5. We propose to use the following finite element spaces on a quadrilateral or prismatic mesh $\mathcal{T}_h$

$$\Sigma_h^k := \begin{cases} \{\tau_h \in L^2_{\text{SYM}}(\Omega) \,:\, \tau_{h,\boldsymbol{x}} \in \Sigma_{\boldsymbol{x}}^{k+1} \otimes \mathcal{P}_z^{k+1}, \\ \quad \tau_{h,\boldsymbol{x}z} \in \mathcal{P}_{\boldsymbol{x}}^k \otimes \mathcal{P}_z^k, \tau_{h,z} \in \mathcal{P}_{\boldsymbol{x}}^{k+1} \otimes \Sigma_z^{k+1}\} & \text{if } d = 2, \\ \{\tau_h \in L^2_{\text{SYM}}(\Omega) \,:\, \tau_{h,\boldsymbol{x}} \in \Sigma_{\boldsymbol{x}}^k \otimes \mathcal{P}_z^{k+1}, \\ \quad \tau_{h,\boldsymbol{x}z} \in \mathcal{P}_{\boldsymbol{x}}^k \otimes \mathcal{P}_z^k, \tau_{h,z} \in \mathcal{P}_{\boldsymbol{x}}^{k+1} \otimes \Sigma_z^{k+1}\} & \text{if } d = 3. \end{cases} \tag{6.9}$$

We note that continuity of $v_{h,\tau}$ and $\tau_{h,nn}$ are satisfied for any $v_h \in V_h^k$ and $\tau_h \in \Sigma_h^k$. Also the Dirichlet boundary conditions $v_{h,\tau} = 0$ on Γ are already included in this definition.

6.1.3 Anisotropic finite elements

In this section, we provide finite element basis functions for quadrilateral and prismatic finite elements. We first define the respective reference elements, and recall some details concerning the mapping Φ_T, which maps the reference element to an element in the mesh. Then, shape functions on the reference elements are given. For the displacements, high order $H(\text{curl})$ conforming shapes as defined in [Zag06] are used. For the stresses, we construct finite element bases using the ones on simplicial elements, which are given in Chapter 4.

6.1.3.1 Reference elements

In the following, let $\hat{T}_1$ be the unit line segment, and $\hat{T}_2$ the triangular reference element, as given in Section 4.4.1. We index these simplices with $\boldsymbol{x}, z$, to demonstrate if they are considered with respect to coordinates $\boldsymbol{x}$ or z, respectively. Thus, in two space dimensions, $\hat{T}_1^{\boldsymbol{x}}, \hat{T}_1^z$ are the reference segments in $\boldsymbol{x}, z$ direction, whereas in three dimensions, we use $\hat{T}_2^{\boldsymbol{x}}, \hat{T}_1^z$ for the in-plane triangular element and the vertical segment.

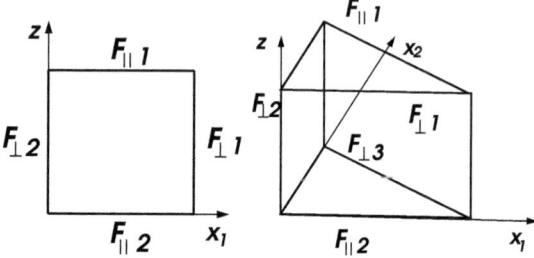

Figure 6.1: Reference quadrilateral and prism

Then, the quadrilateral reference element shall be defined as the tensor product of two line segments, $\hat{T}_2^{x,z} := \hat{T}_1^x \times \hat{T}_1^z$. Analogously, the prismatic reference element is the product of an in-plane triangular element, and a line segment, $\hat{T}_3^{x,z} := \hat{T}_2^x \times \hat{T}_1^z$. We drop the lower index giving the dimension of the respective simplices, in order to unify notation for the two and three dimensional case.

As for the tensor product mesh, we can split the set of facets $\mathcal{F}(\hat{T})$ into two parts, the in-plane facets $\mathcal{F}_\|(\hat{T}) = \{\hat{F}_{\|,1}, \hat{F}_{\|,2}\}$ and the vertical ones $\mathcal{F}_\perp(\hat{T}) = \{\hat{F}_{\perp,i} : i = 1, \ldots, d\}$. In Figure 6.1, we provide a sketch of the quadrilateral and prismatic reference elements.

We first define the notion of polynomial spaces of mixed order on tensor product elements. Let α_1, α_2 be variables of dimensions $d_1, d_2 = 1, 2$. We consider two simplices $T_{d_1}^{\alpha_1}, T_{d_2}^{\alpha_2}$ of dimensions d_1, d_2. Let $T = T_{d_1}^{\alpha_1} \times T_{d_2}^{\alpha_2}$ be their tensor product domain, and let $k_1, k_2 \geq 0$ be integers. Then we define the space of mixed polynomial order (k_1, k_2) with respect to (α_1, α_2) by

$$Q_{\alpha_1,\alpha_2}^{k_1,k_2}(T) := P_{\alpha_1}^{k_1}(T_{d_1}^{\alpha_1}) \otimes P_{\alpha_2}^{k_2}(T_{d_2}^{\alpha_2}).$$

Mainly, we will set $\alpha_1 = \boldsymbol{x}, \alpha_2 = z$. Other choices may occur, for example when describing a quadrilateral facet of the reference prism.

We keep the notion of barycentric coordinates, as they were defined in Section 4.4.1. We use λ_i^x, λ_i^z on $\hat{T}^x, \hat{T}^z$, respectively. There, λ_i^x depends only on $\hat{\boldsymbol{x}}$, is linear, vanishes on facet $\hat{F}_{\perp,i}$ and is one in the edge opposite this facet. This means we set

$$\lambda_1^x = 1 - \hat{x}_1, \qquad \lambda_2^x = \hat{x}_1, \qquad \qquad \text{if } d = 2,$$
$$\lambda_1^x = 1 - \hat{x}_1 - \hat{x}_2, \qquad \lambda_2^x = \hat{x}_1, \qquad \lambda_3^x = \hat{x}_2, \qquad \text{if } d = 3.$$

Also λ_i^z is a linear function only in $\hat{z}$, vanishing on the in-plane facet $\hat{F}_{i,\|}$ and taking value one in the opposite facet,

$$\lambda_1^z = 1 - \hat{z}, \qquad \lambda_2^z = \hat{z}.$$

In Section 4.4.1, we also defined families of polynomials on the reference segment and triangle. In the anisotropic setting, we reuse these families. We shortly recall that

- $\{q_i(\lambda_1, \lambda_2) : 0 \leq i \leq k\}$ span $P^k(\hat{T}_1)$,
- $\{u_i(\lambda_1, \lambda_2) : 2 \leq i \leq k\}$ span $P_0^k(\hat{T}_1)$,
- $\{q_{ij}(\lambda_1, \lambda_2, \lambda_3) : 0 \leq i+j \leq k\}$ span $P^k(\hat{T}_2)$,
- $\{u_i(\lambda_1, \lambda_2)v_j(\lambda_1, \lambda_2, \lambda_3) : 3 \leq i+j \leq k\}$ span $P_0^k(\hat{T}_2)$.

We may use indices such as q_i^x or q_i^z to make clear that the corresponding simplex is $\hat{T}^x$ or $\hat{T}^z$.

Transformation The mapping $\Phi_T : \hat{T} \to T$, which maps the reference element to an element in the mesh, can also be composed by the respective in-plane and transversal mappings. Let now $T = T^x \times T^z$ be an element in the tensor product mesh. Let Φ_{T^x}, Φ_{T^z} be the respective mappings from the reference elements $\hat{T}^x$, $\hat{T}^z$ to T^x, T^z. Then, the compound transformation reads

$$\Phi_T(\hat{x}, \hat{z}) = \begin{pmatrix} \Phi_{T^x}(\hat{x}) \\ \Phi_{T^z}(\hat{z}) \end{pmatrix}.$$

The Jacobian F_T of this map is constant. It is moreover block diagonal,

$$F_T = \begin{pmatrix} F_{T^x} & 0 \\ 0 & F_{T^z} \end{pmatrix}.$$

The Jacobians forming the sub-blocks stem from shape-regular transformations, thus

$$|F_{T^x}|^{-1} \simeq |F_{T^x}^{-1}| \simeq h_x^{-1}, \qquad |F_{T^z}|^{-1} \simeq |F_{T^z}^{-1}| \simeq h_z^{-1}.$$

This allows to introduce

$$\tilde{F}_T := \begin{pmatrix} h_x I & 0 \\ 0 & h_z I \end{pmatrix} \simeq F_T, \qquad (6.10)$$

where I denotes the identity matrix of dimensions one or two. We use these similarities when proving stability and error estimates.

In the following, we construct shape functions for quadrilateral and prismatic elements. In both cases, we proceed in a similar way as for triangular or tetrahedral elements in Sections 4.4.2 and 4.4.3. We define constant tensor fields $\hat{S}^x$, $\hat{S}^z$ and $\hat{S}^{xz}$ which resemble the fields $\hat{S}^{F_i}$ in the simplicial case. Using these, and corresponding bubble fields $\hat{B}^x$, $\hat{B}^z$ and $\hat{B}^{xz}$, we can then construct a finite element basis in a straightforward manner. These constant and bubble fields will play an important role in the stability analysis in Section 6.1.4.

6.1.3.2 On the quadrilateral

In the following, we provide shape functions for the stresses on the quadrilateral reference finite element $\hat{T}$. For the displacements, we refer the interested reader to [Zag06] for a detailed discussion.

We first provide a family of constant tensor fields $\{\hat{S}^x, \hat{S}^z, \hat{S}^{xz}\}$, which span the space of piecewise constant functions. By multiplication with scalar bubble functions $b^x := \lambda_1^x \lambda_2^x$ and $b^z := \lambda_1^z \lambda_2^z$, we then obtain stress bubble tensors $\{\hat{B}^x, \hat{B}^z, \hat{B}^{xz}\}$.

The first pair of fields $(\hat{S}^x, \hat{B}^x)$ is associated with the vertical facets $F \in \mathcal{F}_\perp(\hat{T})$.

$$\hat{S}^x := \begin{pmatrix} 1 & 0 \\ 0 & 0 \end{pmatrix}, \qquad \hat{B}^x := b^x \hat{S}^x.$$

The unit normal of such a facet is $n = \pm e_{x_1}$, therefore $\hat{S}^x_{nn}$ takes value one there, and vanishes on in-plane facets from $\mathcal{F}_\|(\hat{T})$.

Similarly to this field, we define

$$\hat{S}^z := \begin{pmatrix} 0 & 0 \\ 0 & 1 \end{pmatrix}, \qquad \hat{B}^z := b^z \hat{S}^z.$$

They correspond to in-plane facets, as $\hat{S}^z_{nn}$ equals one there.

The third constant tensor field, $\hat{S}^{xz}$ is already a stress bubble function, as its normal-normal component vanishes on all facets

$$\hat{S}^{xz} := \hat{B}^{xz} := \begin{pmatrix} 0 & 1 \\ 1 & 0 \end{pmatrix}.$$

Now, we provide a finite element basis $\hat{\Psi}_k^\Sigma$ for $\hat{\Sigma}^k$. As for the simplicial element types, we organize it in facet- and cell-based functions,

$$\hat{\Psi}_k^\Sigma := \left(\bigcup_{\hat{F}_\| \in \mathcal{F}_\|(\hat{T})} \hat{\Psi}_{\hat{F}_\|,k}^\Sigma \right) \cup \left(\bigcup_{\hat{F}_\perp \in \mathcal{F}_\perp(\hat{T})} \hat{\Psi}_{\hat{F}_\perp,k}^\Sigma \right) \cup \hat{\Psi}_{\hat{T},k}^\Sigma.$$

We use

- the set of shapes corresponding to an in-plane facet $\hat{F}_{\|,m}$, $\hat{\Psi}_{\hat{F}_{\|,m},k}^\Sigma := \{\psi_{\hat{F}_{\|,m},i}^\Sigma\}$

$$\psi_{\hat{F}_{\|,m},i}^\Sigma := q_i^x \hat{S}^z \lambda_{m+1}^z, \qquad 0 \leq i \leq k+1,$$

- the set of shapes corresponding to a vertical facet $\hat{F}_{\perp,m}$, $\hat{\Psi}_{\hat{F}_{\perp,m},k}^\Sigma := \{\psi_{\hat{F}_{\perp,m},i}^\Sigma\}$

$$\psi_{\hat{F}_{\perp,m},i}^\Sigma := q_i^z \hat{S}^x \lambda_{m+1}^x, \qquad 0 \leq i \leq k+1,$$

- the set of cell-based shapes $\hat{\Psi}^{\Sigma}_{\hat{T},k} := \{\psi^{\Sigma}_{\hat{T},x,ij}, \psi^{\Sigma}_{\hat{T},z,ij}, \psi^{\Sigma}_{\hat{T},xz,ij}\}$

$$\psi^{\Sigma}_{\hat{T},x,ij} := q^x_i q^z_j \hat{B}^x \qquad 0 \leq i \leq k-1,\ 0 \leq j \leq k+1,$$
$$\psi^{\Sigma}_{\hat{T},z,ij} := q^x_i q^z_j \hat{B}^z \qquad 0 \leq i \leq k+1,\ 0 \leq j \leq k-1,$$
$$\psi^{\Sigma}_{\hat{T},xz,ij} := q^x_i q^z_j \hat{B}^{xz} \qquad 0 \leq i \leq k,\ 0 \leq j \leq k.$$

Here, the index of the barycentric coordinates is to be seen modulo 2. The construction ensures that the normal-normal component of a shape from $\hat{\Psi}^{\Sigma}_{\hat{F},k}$ takes values only on facet $\hat{F}$ and not on the other facets. We observe that the normal-normal components of the facet-based shapes span $P^{k+1}(\hat{F})$ for all facets $\hat{F} \in \mathcal{F}(\hat{T})$. One can easily check the linear independence of the basis functions. By counting the degrees of freedom for the respective components, one can see that the proposed shape functions span the local polynomial space suggested in the definition of the quadrilateral finite element space (6.9).

6.1.3.3 On the prism

On the prismatic reference element $\hat{T}$, we proceed in a similar manner as for the quadrilateral element. To define the family of constant tensor fields, we recall the two-by-two symmetric tensor fields $\hat{S}^{\hat{F}_i}$ defined on the triangle, see equation (4.46). We set, using the scalar function $b^x_i := \lambda^x_i$

$$\hat{S}^x_i := \begin{pmatrix} \hat{S}^{\hat{F}_i} & 0 \\ 0 & 0 \end{pmatrix}, \qquad \hat{B}^x_i := b^x_i \hat{S}^x_i, \qquad i = 1, 2, 3.$$

Then, $\hat{S}^x_{i,nn}$ vanishes on all facets but $\hat{F}_{\perp,i}$. As b^x_i it is zero on facet $F_{\perp,i}$, the tensor field $\hat{B}^x_{i,nn}$ takes zero values on the whole boundary $\partial \hat{T}$. For the upper and lower triangular facet, we define a constant field $\hat{S}^z$ whose normal-normal component takes value one there, and vanishes on all other facets. Then we obtain a bubble by multiplication with $b^z = \lambda^z_1 \lambda^z_2$,

$$\hat{S}^z := \begin{pmatrix} 0 & 0 \\ 0 & 1 \end{pmatrix}, \qquad \hat{B}^z := b^z \hat{S}^z.$$

There remain two more fields, which are needed to span the space of constant, symmetric tensor fields on the prism. We define these fields $\hat{S}^{xz}_i$, $i = 1, 2$ such that they are already a stress bubble function, as we did in the two-dimensional case

$$\hat{S}^{xz}_i := \hat{B}^{xz}_i := \begin{pmatrix} 0 & e_{x_i} \\ e^T_{x_i} & 0 \end{pmatrix}, \qquad i = 1, 2.$$

From this framework, we can again define a finite element basis $\hat{\Psi}^{\Sigma}_k$ for $\hat{\Sigma}^k$ on the reference prism. It is again divided into facet- and cell-based functions,

$$\hat{\Psi}_k^\Sigma := \left(\bigcup_{\hat{F}_\| \in \mathcal{F}_\|(\hat{T})} \hat{\Psi}_{\hat{F}_\|,k}^\Sigma\right) \cup \left(\bigcup_{\hat{F}_\perp \in \mathcal{F}_\perp(\hat{T})} \hat{\Psi}_{\hat{F}_\perp,k}^\Sigma\right) \cup \hat{\Psi}_{\hat{T},k}^\Sigma.$$

Here we use

- the set of shapes corresponding to an in-plane facet $\hat{F}_{\|,m}$, $\hat{\Psi}_{\hat{F}_{\|,m},k}^\Sigma := \{\psi_{\hat{F}_{\|,m},ij}^\Sigma\}$

$$\psi_{\hat{F}_{\|,m},ij}^\Sigma := q_{ij}^x \hat{S}^z \lambda_{m+1}^z, \qquad 0 \le i+j \le k+1,$$

- the set of shapes corresponding to a vertical facet $\hat{F}_{\perp,m}$, $\hat{\Psi}_{\hat{F}_{\perp,m},k}^\Sigma := \{\psi_{\hat{F}_{\perp,m},ij}^\Sigma\}$

$$\psi_{\hat{F}_{\perp,m},ij}^\Sigma := q_{i0}^x q_j^z \hat{S}_m^x, \qquad 0 \le i \le k, \ 0 \le j \le k+1,$$

- the set of cell-based shapes $\hat{\Psi}_{\hat{T},k}^\Sigma := \{\psi_{\hat{T},x,ijl}^\Sigma, \psi_{\hat{T},z,ijl}^\Sigma, \psi_{\hat{T},xz,ijl}^\Sigma\}$

$$\begin{aligned}
\psi_{\hat{T},x,ijl}^\Sigma &:= q_{ij}^x q_l^z \hat{B}^x & 0 \le i+j \le k-1, \ 0 \le l \le k+1, \\
\psi_{\hat{T},z,ijl}^\Sigma &:= q_{ij}^x q_l^z \hat{B}^z & 0 \le i+j \le k+1, \ 0 \le l \le k-1, \\
\psi_{\hat{T},xz,ijl}^\Sigma &:= q_{ij}^x q_l^z \hat{B}^{xz} & 0 \le i+j \le k, \ 0 \le l \le k.
\end{aligned}$$

As for the case of quadrilaterals, one can see that the normal-normal component of a facet-based shape from $\hat{\Psi}_{\hat{F},k}^\Sigma$ vanishes on all facets but $\hat{F}$. On a triangular facet $\hat{F}_\| \in \mathcal{F}_\|(\hat{T})$, the normal-normal components of $\hat{\Psi}_{\hat{F}_\|}^\Sigma$ span $P^{k+1}(\hat{F}_\|)$. For a quadrilateral facet $\hat{F}_\perp \in \mathcal{F}_\perp(\hat{T})$, we obtain $Q^{k,k+1}(\hat{F}_\perp)$ instead. Linear independence of the basis can be seen directly. The proposed set of shape functions spans the local polynomial space suggested in (6.9) for a finite element space on a prismatic mesh.

6.1.3.4 Global finite element spaces

The basis functions given on the reference quadrilateral or prismatic element are transformed to any element $T \in \mathcal{T}_h$ by the conforming transformation Φ_T^Σ, which is given in Section 4.2.2.5. Then, one obtains the global finite element space Σ_h^k. As for simplicial elements, the local space can be split into a facet- and a cell-bound part, $\Sigma^k(\hat{T}) = \Sigma^{k,f}(\hat{T}) + \Sigma^{k,b}(\hat{T})$. There, the subspaces are spanned by the facet and inner basis functions, respectively. This splitting can analogously be performed for the global finite element space, $\Sigma_h^k = \Sigma_h^{k,f} + \Sigma_h^{k,b}$.

6.1.4 Stability for tensor-product elements

In this section, we concentrate on stability estimates for the discrete mixed problem in the anisotropic setup.

Problem 6.2 (Elasticity on a tensor product domain)**.** *Let Ω be a tensor product domain according to Definition 6.1. Let $\mathcal{T}_h$ be the corresponding quadrilateral or prismatic triangulation. Find $(\sigma_h, u_h) \in \Sigma_h^k \times V_h^k$ such that*

$$\begin{aligned} a(\sigma_h, \tau_h) &+ b(\tau_h, u_h) &= \langle F_1, \tau_h \rangle_\Sigma & \quad \forall \tau_h \in \Sigma_h^k, \\ b(\sigma_h, v_h) & &= \langle F_2, v_h \rangle_V & \quad \forall v_h \in V_h^k. \end{aligned} \tag{6.11}$$

We emphasize that all stability estimates derived in the following do neither depend on the anisotropic mesh sizes h_x, h_z nor their ratio h_x/h_z. In our analysis, we use the discrete norms $\|\cdot\|_{\Sigma_h}$ and $\|\cdot\|_{V_h}$ proposed in Chapter 4,

$$\begin{aligned} \|v_h\|_{V_h}^2 &:= \sum_{T \in \mathcal{T}_h} \|\varepsilon(v_h)\|_T^2 + \sum_{F \in \mathcal{F}_h} h_F^{-1} \|[\![v_h]\!]_n\|_F^2, \\ \|\tau_h\|_{\Sigma_h} &:= \|\tau_h\|_\Omega. \end{aligned}$$

Our theory relies on the fact that only piecewise strains, and not gradients are used for the displacement norm. Beforehand, we show a norm equivalence on the discrete stress space. In Lemma 4.14, we already proved such an estimate for shape-regular meshes. Now we need to take the anisotropic structure into account.

Lemma 6.3. *Let $T \in \mathcal{T}_h$ be a quadrilateral or prismatic element in the triangulation $\mathcal{T}_h$.*

1. *For $\tau_h \in \Sigma^k(T)$, there holds the trace inequality*

$$\sum_{F \in \mathcal{F}(T)} h_F \|\tau_{h,nn}\|_F^2 \leq c \|\tau_h\|_T^2. \tag{6.12}$$

2. *Let g be in the normal-normal trace space of $\Sigma^k(T)$, i.e. $g \in P_x^{k+1}(F_\|)$ for an in-plane facet $F_\|$, $g \in P^{k+1}(F_\perp)$ for a vertical facet and $d = 2$, or $g \in Q^{k,k+1}(F_\perp)$ for a vertical facet and $d = 3$. Then there exists an extension $\tau_h \in \Sigma^k(T)$ such that $\tau_{h,nn} = g$ on ∂T, and*

$$\|\tau_h\|_T^2 \leq c \sum_{F \in \mathcal{F}(T)} h_F \|g\|_F^2. \tag{6.13}$$

The constants are independent of the anisotropic mesh sizes h_x, h_z, and their ratio h_x/h_z.

Proof. 1. We decompose the finite element function into a facet- and a cell-based part,

$$\tau_h = \tau_h^f + \tau_h^b,$$

where $\tau_h^f \in \Sigma^{k,f}(T)$ and $\tau_h^b \in \Sigma^{k,b}(T)$. There, the facet-based function τ_h^f can be further split into the contributions coming from the different facets,

$$\tau_h^f = \sum_{F_\| \in \mathcal{F}_\|(T)} \tau_h^{F_\|} + \sum_{F_\perp \in \mathcal{F}_\perp(T)} \tau_h^{F_\perp},$$

where τ_h^F is built from shape functions matching facet F. As the basis functions are linearly independent, and the local space is of finite, bounded dimensions, we may estimate

$$\|\tau_h\|_T \geq c \left(\sum_{F \in \mathcal{F}(T)} \|\tau_h^F\|_T^2 + \|\tau_h^b\|_T^2 \right).$$

Let now $F_{\|,m} \in \mathcal{F}_{\|}(T)$ be an in-plane facet, and $F_\perp \in \mathcal{F}_{\perp,m}(T)$ be a facet in transversal direction. From the construction of the shape functions in Section 6.1.3, we know that $\tau_h^{F_\|}, \tau_h^{F_\perp}$ are of the special form

$$\tau_h^{F_{\|,m}} = \frac{1}{J_T^2} F_T \hat{\tau}_h^{F_{\|,m}} F_T^T, \qquad \hat{\tau}_h^{F_{\|,m}} = q_m^{\boldsymbol{x}} \lambda_{m+1}^z \hat{S}^z,$$

$$\tau_h^{F_{\perp,m}} = \frac{1}{J_T^2} F_T \hat{\tau}_h^{F_{\perp,m}} F_T^T, \qquad \hat{\tau}_h^{F_{\perp,m}} = \begin{cases} q_m^z \lambda_{m+1}^{\boldsymbol{x}} \hat{S}^{\boldsymbol{x}} & \text{if } d = 2, \\ q_m^z \hat{S}_m^{\boldsymbol{x}} & \text{if } d = 3. \end{cases}$$

where $q_m^{\boldsymbol{x}}, q_m^z$ are scalar polynomials with respect to $\boldsymbol{x}, z$. Using the tensor product structure of $\hat{S}^{\boldsymbol{x}}, \hat{S}^z$, and the fact that the Jacobian F_T is equivalent to the block-diagonal matrix $\tilde{F}_T$ from equation (6.10) we see, for the in-plane facet case

$$\left| \sum_{m=1}^2 \tau_h^{F_{\|,m}} \right|^2 \simeq \left| \sum_{m=1}^2 \frac{q_m^{\boldsymbol{x}} \lambda_{m+1}^z}{J_T^2} \tilde{F}_T \hat{S}^z \tilde{F}_T^T \right|^2 \simeq \frac{h_z^4}{J_T^4} \left| \sum_{m=1}^2 q_m^{\boldsymbol{x}} \lambda_{m+1}^z \right|^2 \simeq \frac{h_z^4}{J_T^4} \left| \sum_{m=1}^2 \hat{\tau}^{F_{\|,m}} \right|^2.$$

For the transversal direction, we differ between two and three space dimensions. In two dimensions, we obtain

$$\left| \sum_{m=1}^2 \tau_h^{F_{\perp,m}} \right|^2 \simeq \frac{h_{\boldsymbol{x}}^4}{J_T^4} \left| \sum_{m=1}^2 \hat{\tau}_h^{F_{\perp,m}} \right|^2$$

by exchanging z with $\boldsymbol{x}$ and $\|$ with $\perp$ in the estimate above. For the case of three dimensions, we see

$$\left| \sum_{m=1}^3 \tau_h^{F_{\perp,m}} \right|^2 \simeq \left| \sum_{m=1}^3 \frac{q_m^{\boldsymbol{x}}}{J_T^2} \tilde{F}_T \hat{S}_m^{\boldsymbol{x}} \tilde{F}_T^T \right|^2 \simeq \frac{h_{\boldsymbol{x}}^4}{J_T^4} \left| \sum_{m=1}^3 q_m^{\boldsymbol{x}} \hat{S}_m^{\boldsymbol{x}} \right|^2 = \frac{h_{\boldsymbol{x}}^4}{J_T^4} \left| \sum_{m=1}^3 \hat{\tau}_h^{F_{\perp,m}} \right|^2,$$

where we have used that the $\hat{S}_i^{\boldsymbol{x}}$ are linearly independent. These estimates together imply that

$$\left| \sum_{F \in \mathcal{F}(T)} \tau_h^F \right|^2 \simeq \frac{h_F^4}{J_T^4} \left| \sum_{F \in \mathcal{F}(T)} \hat{\tau}_h^F \right|^2.$$

Note that we have used the shape-regularity of the in-plane and transversal meshes, from which follows $h_x \simeq h_{F_\perp}$ and $h_z \simeq h_{F_\parallel}$. We proceed, using $h_F \simeq J_T/J_F$,

$$\begin{aligned}
\|\tau_h^f\|_T^2 &= \int_T \left| \sum_{F \in \mathcal{F}(T)} \tau_h^F \right|^2 dx \simeq \int_{\hat{T}} \frac{h_F^4}{J_T^4} \left| \sum_{F \in \mathcal{F}(T)} \hat{\tau}_h^F \right|^2 J_T \, d\hat{x} \\
&\simeq \sum_{F \in \mathcal{F}(T)} \int_{\hat{F}} \frac{h_F^4}{J_T^3} |\hat{\tau}_{h,nn}^F|^2 \, d\hat{s} = \sum_{F \in \mathcal{F}(T)} \int_F \frac{h_F^4}{J_T^3} J_F^4 |\hat{\tau}_{h,nn}^F|^2 \frac{1}{J_F} \, ds \\
&\simeq \sum_{F \in \mathcal{F}(T)} h_F \|\tau_{h,nn}^F\|_F^2 = \sum_{F \in \mathcal{F}(T)} h_F \|\tau_{h,nn}\|_F^2.
\end{aligned}$$

This estimate implies the trace inequality (6.12). We will moreover use it in the second part of the proof.

2. Due to the construction of the subspace $\Sigma_h^{k,f}$, it is possible to find a unique extension $\tau_h^f \in \Sigma_h^{k,f}(T)$ of g. The function τ_h^f is the required extension due to the equivalence

$$\|\tau_h^f\|_T^2 \simeq \sum_{F \in \mathcal{F}(T)} h_F^{-1} \|\tau_{h,nn}\|_F^2.$$

□

Corollary 6.4. *There exists a constant $c > 0$ independent of h_x, h_z or their ratio h_x/h_z, such that for $\tau_h \in \Sigma_h^k$*

$$\|\tau_h\|_{\Sigma_h}^2 \leq \|\tau_h\|_\Omega^2 + \sum_{F \in \mathcal{F}_h} h_F \|\tau_{h,nn}\|_F^2 \leq c\|\tau_h\|_{\Sigma_h}^2.$$

We finally show inf-sup stability of $b(\cdot,\cdot)$ on $\Sigma_h^k \times V_h^k$ with respect to the discrete norms. The proof for the following lemma runs along the same lines as the one for Lemma 4.17. Therefore, we do not repeat it in detail, but only comment on the independence of the stability constant of the anisotropic mesh sizes h_x, h_z or their ratio h_x/h_z.

Lemma 6.5. *Let Ω be a tensor product domain according to Definition 6.1. Let $\mathcal{T}_h$ be the corresponding tensor-product triangulation, stemming from shape-regular, quasi-uniform triangulations $\mathcal{T}_{h_x}^x, \mathcal{T}_{h_z}^z$. There holds the stability estimate*

$$\inf_{v_h \in V_h^k} \sup_{\tau_h \in \Sigma_h^k} \frac{b(\tau_h, v_h)}{\|\tau_h\|_{\Sigma_h} \|v_h\|_{V_h}} \geq \tilde{c}_{b,1},$$

where $\tilde{c}_{b,1} > 0$ is independent of the anisotropic mesh sizes h_x, h_z or their ratio h_x/h_z.

Proof. We recall the proof of the inf-sup condition on simplicial, shape-regular meshes (see Chapter 4, Lemma 4.17). We reuse these ideas, and construct a finite element function $\tau_h = c_f \tau_h^f + c_b \tau_h^b$, which consists of a facet part $\tau_h^f \in \Sigma_h^{k,f}$, and a bubble part $\tau_h^b \in \Sigma_h^{k,b}$. There, τ_h^f can still be chosen such that

$$\tau_{h,nn}^f |_F = h_F^{-1} [\![v_h]\!]_{n,F} \quad \forall F \in \mathcal{F}_h,$$

as one can check comparing the respective polynomial orders. We define the bubble part τ_h^b element-wise, such that $\tau_h^b|_T := \Phi_T \hat{\tau}_h^b$. Here, we set $\hat{\tau}_h^b$ on the reference element,

$$\hat{\tau}_h^b := J_T^2 \sum_m \left(\hat{\varepsilon}(\hat{v}_h) : \tilde{F}_T^{-T} \hat{S}_m \tilde{F}_T^{-1} \right) \tilde{F}_T^{-T} \hat{B}_m \tilde{F}_T^{-1},$$

where $\{\hat{S}_m\}$, $\{\hat{B}_m\}$ are the respective unions of the fields $\hat{S}_m^x, \hat{S}_n^{xz}, \hat{S}^z$ and $\hat{B}_m^x, \hat{B}_n^{xz}, \hat{B}^z$, and $\tilde{F}_T \simeq F_T$ is taken from equation (6.10). Due to the diagonal block structure of $\tilde{F}$, this tensor is a bubble function. For these choices, all estimates from the proof of Lemma 4.17 work the same way, but the equivalence $|F_T^{-1}|_s \simeq h_T$ is replaced by $|F_T^{-1}|_s \simeq |\tilde{F}_T^{-1}|_s$. As this holds independently of the aspect ratio of the finite elements, we arrive at the required results. □

6.2 Interpolation operators

In this section, we provide interpolation operators for the spaces Σ_h^k, V_h^k on the tensor product mesh $\mathcal{T}_h$. For the displacement space, we use a Clément-type interpolation operator $\mathcal{Q}_{h,k}^V$, which is built from quasi-interpolation operators for one and two dimensions proposed in [Sch01]. We provide an interpolation error estimate in the discrete norm $\|\cdot\|_{V_h}$, which does not depend on the aspect ratio h_x/h_z of the elements. There, we use the commuting diagram property of the family of operators from [Sch01], and their L^2 stability. For the stresses, we propose to use a tensor product operator $\mathcal{I}_{h,k}^\Sigma$ built from nodal interpolation operators for the normal-normal continuous and non-continuous spaces in the plane and on the line.

We first recall basic properties of the nodal interpolation operators

$$\begin{aligned}\mathcal{I}_{x,k}^\Sigma &: H(\mathrm{div\,div}; \Omega_x) \to \Sigma_x^k, & \mathcal{I}_{z,k}^\Sigma &: H(\mathrm{div\,div}; \Omega_z) \to \Sigma_z^k, & (6.14) \\ \mathcal{I}_{x,k}^\mathcal{P} &: L^2(\Omega_x) \to \mathcal{P}_x^k, & \mathcal{I}_{z,k}^\mathcal{P} &: L^2(\Omega_z) \to \mathcal{P}_z^k. & (6.15)\end{aligned}$$

The interpolation operator $\mathcal{I}_{x,k}^\Sigma$ was defined and analyzed in the scope of Section 4.3.2.3. As space Σ_z^k coincides with the Lagrange space W_z^k, we use the nodal interpolation operator for H^1 for $\mathcal{I}_{z,k}^\Sigma$. For the non-continuous spaces, the interpolators $\mathcal{I}_{x,k}^\mathcal{P}, \mathcal{I}_{z,k}^\mathcal{P}$ are element-wise L^2 projections onto P^k. Thus, estimates for the interpolation error in the L^2 norm for the different operators may be given straight away, we collect them in the following lemma.

Lemma 6.6. *Let $\mathcal{T}_{h_\alpha}^\alpha$ be a shape-regular, quasi-uniform triangulation of Ω_α, $\alpha = x, z$. For $\tau \in H_{\mathrm{SYM}}^m(\Omega_\alpha)$ and $q \in H^m(\Omega_\alpha)$, where $1 \leq m \leq k+1$, the interpolation error for the nodal interpolation operators $\mathcal{I}_{\alpha,k}^\Sigma$ and $\mathcal{I}_{\alpha,k}^\mathcal{P}$ satisfies*

$$\begin{aligned}\|\tau - \mathcal{I}_{\alpha,k}^\Sigma \tau\|_\Omega &\leq c h_\alpha^m |\tau|_{H^m(\Omega_\alpha)}, & (6.16) \\ \|q - \mathcal{I}_{\alpha,k}^\mathcal{P} q\|_\Omega &\leq c h_\alpha^m |q|_{H^m(\Omega_\alpha)}. & (6.17)\end{aligned}$$

The constant $c > 0$ does not depend on the mesh size h_α, but only on the shape regularity of $\mathcal{T}_{h_\alpha}^\alpha$.

6.2.1 Commuting diagram quasi-interpolation operators

In [Sch01], a family of quasi-interpolation operators for H^1, $H(\text{curl})$, $H(\text{div})$ and L^2 were introduced. In our analysis, we need these operators for the H^1- and $H(\text{curl})$-conforming spaces,

$$\mathcal{Q}_{x,k}^W : L^2(\Omega_x) \to W_x^k, \qquad \mathcal{Q}_{z,k}^W : L^2(\Omega_z) \to W_z^k, \qquad (6.18)$$
$$\mathcal{Q}_{x,k}^V : L^2(\Omega_x) \to V_x^k, \qquad \mathcal{Q}_{z,k}^V : L^2(\Omega_z) \to V_z^k. \qquad (6.19)$$

The theory presented in [Sch01] is restricted to three space dimensions, and the lowest order case, i.e. the space of piecewise linear, continuous functions W_h^1 and the Nédélec type I space $V_{h,0}$. Corresponding operators for the one-dimensional case can be derived straightforward, we will not dwell on their construction. From the construction of the low-order operators, one can directly see how to extend these ideas to the high-order case in two and three space dimensions. Basically, the degrees of freedom of the finite elements are replaced by weighted local averages, where the weighting functions are chosen such that these averages coincide with the nodal values for polynomials.

6.2.1.1 Commuting diagram quasi-interpolation operators in two dimensions

In the following, we provide a sound definition of the interpolation operators $\mathcal{Q}_{x,k}^W$ and $\mathcal{Q}_{x,k}^V$, given the dimension of x is two. To this end, we assume that $T = [V_1, V_2, V_3]$ is a triangle in the two-dimensional mesh $\mathcal{T}_{h_x}^x$. We will drop x as an index in this section, well aware that the theory below will be applied to the x-component within the tensor-product setup. Note that, in two space dimensions, the sets of facets and edges coincide.

To define the quasi-interpolation operators $\mathcal{Q}_{x,k}^W$, $\mathcal{Q}_{x,k}^V$, we first recall the respective nodal operators, as introduced in Section 4.3.2. Therefore, we need the finite elements $(T, W^k(T), \mathcal{N}_k^W(T))$ for H^1 and $(T, V^k(T), \mathcal{N}_k^V(T))$ for $H(\text{curl})$ introduced in Section 4.2.2.1 and Section 4.2.2.2. There, the sets of degrees of freedom were composed by

$$\mathcal{N}_k^W(T) := \left(\bigcup_{V \in \mathcal{V}(T)} \mathcal{N}_V^W\right) \cup \left(\bigcup_{F \in \mathcal{F}(T)} \mathcal{N}_{F,k}^W\right) \cup \mathcal{N}_{T,k}^W, \qquad (6.20)$$

$$\mathcal{N}_k^V(T) := \left(\bigcup_{F \in \mathcal{F}(T)} \mathcal{N}_{F,k}^V\right) \cup \mathcal{N}_{T,k}^V. \qquad (6.21)$$

We assume now that the sets

$$\Psi_k^W := \{\psi_V^W : V \in \mathcal{V}(T)\} \cup \{\psi_{F,i}^W : F \in \mathcal{F}(T)\} \cup \{\psi_{T,i}^W\}, \qquad (6.22)$$
$$\Psi_k^V := \{\psi_{F,i}^W : F \in \mathcal{F}(T)\} \cup \{\psi_{T,i}^W\}, \qquad (6.23)$$

are the corresponding nodal bases, such that the nodal interpolation operators are defined by

$$\mathcal{I}_k^W w := \sum_{V \in \mathcal{V}_{h_x}^x} N_V^W(w) \psi_V^W + \sum_{F \in \mathcal{F}_{h_x}^x} \sum_i N_{F,i}^W(w) \psi_{F,i}^W + \sum_{T \in \mathcal{T}_{h_x}^x} \sum_i N_{T,i}^W(w) \psi_{T,i}^W,$$

$$\mathcal{I}_k^V v := \sum_{F \in \mathcal{F}_{h_x}^x} \sum_i N_{F,i}^V(v) \psi_{F,i}^V + \sum_{T \in \mathcal{T}_{h_x}^x} N_{T,i}^V(v) \psi_{T,i}^V.$$

We recall that the major drawback of these operators is the fact that they are only defined for sufficiently smooth functions, and not stable with respect to the L^2 norm.

The quasi-interpolation operators are derived from the nodal ones, where point evaluations are replaced by local smoothing operators. Facet- and domain-integrals are adapted in a similar way. Following [Sch01], we define for each vertex $V \in \mathcal{V}_{h_x}^x$ a set of non-zero measure $\omega_V \subset \Omega_x \cap \Delta_V$. Next, we define functions $f_V \in L^\infty(\omega_V)$ such that

$$\int_{\omega_V} f_V \, q \, dx = q(V) \quad \forall q \in P^{2k+2}(\Omega_x), \, \forall V \in \mathcal{V}_{h_x}^x, \tag{6.24}$$

$$\|f_V\|_{L^\infty} \simeq h_x^{-2}. \tag{6.25}$$

We now define sets of functionals $\boldsymbol{Q}_k^W(T) := \{Q_V^W, Q_{F,i}^W, Q_{T,i}^W\}$ resembling $\mathcal{N}_k^W(T)$, which are well defined on L^2. Similarly, we introduce degrees of freedom $\boldsymbol{Q}_k^V(T) := \{Q_{F,i}^V, Q_{T,i}^V\}$ for $H(\text{curl})$. We set

- for $V \in \mathcal{V}(T)$,

$$Q_V^W(w) := \int_{\omega_V} f_V(y) \, w(y) \, dy,$$

- for a facet $F \in \mathcal{F}(T)$, $F = [V_1, V_2]$,

$$Q_{F,i}^W(w) := \int_{\omega_{V_1}} \int_{\omega_{V_2}} f_{V_1}(y_1) f_{V_2}(y_2) \int_{[y_1, y_2]} \frac{\partial w}{\partial s} \frac{\partial q_i}{\partial s} \, ds \, dy_2 dy_1$$

with $\{q_i : 2 \leq i \leq k\}$ a basis for $P_0^k(E)$, which is transformed to the facet $[y_1, y_2]$ such that $\partial q_i / \partial s$ is preserved,

- for $T = [V_1, V_2, V_3]$,

$$Q_{T,i}^W(w) := \int_{\omega_{V_1}} \int_{\omega_{V_2}} \int_{\omega_{V_3}} f_{V_1}(y_1) f_{V_2}(y_2) f_{V_3}(y_3) \int_{[y_1, y_2, y_3]} \nabla w \cdot \nabla q_i \, dx \, dy_3 dy_2 dy_1,$$

where $\{q_i\}$ form a basis for $P_0^k(T)$, which is transformed to the triangle $[y_1, y_2, y_3]$ such that ∇q_i is preserved.

For the $H(\text{curl})$ conforming space, we define

- for a facet $F \in \mathcal{F}(T)$, $F = [V_1, V_2]$,

$$Q^V_{F,i}(v) := \int\int_{\omega_{V_1}\,\omega_{V_2}} f_{V_1}(y_1) f_{V_2}(y_2) \int_{[y_1,y_2]} v_\tau\, q_i\, ds\, dy_2 dy_1$$

with $\{q_i : 0 \leq i \leq k\}$ a basis for $P^k(E)$, which is transformed to the facet $[y_1, y_2]$ such that q_i is preserved,

- for $T = [V_1, V_2, V_3]$

$$Q^V_{T,i}(v) := \int\int\int_{\omega_{V_1}\,\omega_{V_2}\,\omega_{V_3}} f_{V_1}(y_1) f_{V_2}(y_2) f_{V_3}(y_3) \int_{[y_1,y_2,y_3]} \operatorname{curl}(v) \cdot \operatorname{curl}(q_i)\, dx\, dy_3 dy_2 dy_1,$$

$$Q^V_{T,j}(v) := \int\int\int_{\omega_{V_1}\,\omega_{V_2}\,\omega_{V_3}} f_{V_1}(y_1) f_{V_2}(y_2) f_{V_3}(y_3) \int_{[y_1,y_2,y_3]} v \cdot r_j\, dx\, dy_3 dy_2 dy_1$$

where $\{q_i\}$ are such that $\{\operatorname{curl} q_i\}$ form a basis for $\operatorname{curl}(P^k_{0,\tau}(T))$, and $\{r_j\}$ are a basis for $\nabla(P^{k+1}_0(T))$, which are transformed to the triangle $[y_1, y_2, y_3]$ such that $\operatorname{curl} q_i, r_j$ are preserved.

Using these smoothed nodal values, the quasi-interpolation operators $\mathcal{Q}^W_{x,k}, \mathcal{Q}^V_{x,k}$ are defined by

$$\mathcal{Q}^W_{x,k} w := \sum_{Q^W_i \in \boldsymbol{Q}^W_k} Q^W_i(w) \psi^W_i, \tag{6.26}$$

$$\mathcal{Q}^V_{x,k} v := \sum_{Q^V_i \in \boldsymbol{Q}^V_k} Q^V_i(v) \psi^V_i, \tag{6.27}$$

where the shape functions ψ^W_i, ψ^V_i are matching the respective degrees of freedom. These operators are consistent in the sense that they preserve polynomials on patches.

Lemma 6.7. *Let the functions f_V satisfy (6.24). The quasi-interpolation operators $\mathcal{Q}^W_{x,k}, \mathcal{Q}^V_{x,k}$ preserve polynomials up to order k on a patch Δ_T, i.e. for $w \in P^k(\Delta_T), v \in [P^k(\Delta_T)]^2$*

$$w = \mathcal{Q}^W_{x,k} w \quad \text{and} \quad v = \mathcal{Q}^V_{x,k} v \quad \text{on } T. \tag{6.28}$$

Proof. We show that for polynomial w, v, the quasi-interpolation operators coincide with the nodal ones, which preserve polynomials by definition. This means, we have to ensure that, for all matching pairs of degrees of freedom $(N^W, Q^W) \in \mathcal{N}^W_k \times \boldsymbol{Q}^W_k$ and $(N^V, Q^V) \in \mathcal{N}^V_k \times \boldsymbol{Q}^V_k$,

$$N^W(w) = Q^W(w), \quad \text{and} \quad N^V(v) = Q^V(v).$$

This was done for the vertex degrees of freedom N^W_V and the lowest order edge degrees of freedom $N^V_{F,0}$ in [Sch01]. The respective equivalences for high-order facet degrees of freedom for both H^1

and $H(\text{curl})$ follow along the same line. We present the proof on facet $F = [V_1, V_2]$ for the $H(\text{curl})$ case. We use a transformation of the facet $[y_1, y_2]$ to the reference facet $\hat{F} = [0, 1]$, and the fact that f_V preserves point values for polynomials according to equation (6.24). We set $s := y_1 + \hat{s}(y_2 - y_1)$, and $\hat{v}_{\hat{\tau}}(\hat{s}) := v_\tau(s), \hat{q}_i(\hat{s}) := q_i(s)$. Note that $\hat{v}_{\hat{\tau}}(\hat{s})\hat{q}_i(\hat{s})J_{[y_1,y_2]}$ lies in P^{2k+1} with respect to both y_1, y_2 separately.

$$\begin{aligned}
Q_{F,i}^V(v) &= \int_{\omega_{V_1}} \int_{\omega_{V_2}} f_{V_1}(y_1) f_{V_2}(y_2) \int_{[y_1,y_2]} v_\tau \cdot q_i \, ds \, dy_2 dy_1 \\
&= \int_{\hat{F}} \int_{\omega_{V_1}} f_{V_1}(y_1) \int_{\omega_{V_2}} f_{V_2}(y_2) \underbrace{\hat{v}_{\hat{\tau}}(\hat{s}) \cdot \hat{q}_i(\hat{s}) J_{[y_1,y_2]}}_{\in P_{y_2}^{2k+1}} dy_2 dy_1 \, d\hat{s} \\
&= \int_{\hat{F}} \int_{\omega_{V_1}} f_{V_1}(y_1) \underbrace{\hat{v}_{\hat{\tau}}(\hat{s}) \cdot \hat{q}_i(\hat{s}) J_{[y_1,V_2]}}_{\in P_{y_1}^{2k+1}} dy_1 \, d\hat{s} \\
&= \int_{\hat{F}} \hat{v}_{\hat{\tau}} \cdot \hat{q}_i J_F \, d\hat{s} = \int_F v_\tau \cdot q_i \, ds.
\end{aligned}$$

It remains to prove the equality for the cell-based degrees of freedom. We first show

$$N_{T,j}^V(v) = Q_{T,j}^V(v),$$

a similar proof holds then for the cell-based H^1 conforming degrees of freedom. Again, we perform a transformation to the reference element, such that $\hat{x}$ corresponds to the point x in the triangle $[y_1, y_2, y_3]$. Then, setting $\hat{v}(\hat{x}) := v(x), \hat{r}_j(\hat{x}) := r_j(x)$, we see that the polynomial $\hat{v}(\hat{x}) \cdot \hat{r}_j(\hat{x}) J_{[y_1,y_2,y_3]}$ lies in P^{2k+2} with respect to the y_i. This ensures

$$\begin{aligned}
Q_{T,i}^V(v) &= \int_{\omega_{V_1}} \int_{\omega_{V_2}} \int_{\omega_{V_3}} f_{V_1}(y_1) f_{V_2}(y_2) f_{V_3}(y_3) \int_{[y_1,y_2,y_3]} v \cdot r_j \, dx \, dy_3 dy_2 dy_1 \\
&= \int_{\hat{T}} \int_{\omega_{V_1}} f_{f_1}(y_1) \int_{\omega_{V_2}} f_{V_2}(y_2) \int_{\omega_{V_3}} f_{V_3}(y_3) \underbrace{\hat{v}(\hat{x}) \cdot \hat{r}_j(\hat{x}) J_{[y_1,y_2,y_3]}}_{\in P^{2k+2}} dy_3 dy_2 dy_1 \, d\hat{x} \\
&= \int_{\hat{T}} \hat{v}(\hat{x}) \cdot \hat{r}_j(\hat{x}) J_T \, d\hat{x} = \int_T v \cdot r_j \, dx.
\end{aligned}$$

Last, we show the equality $N_{T,i}^V(v) = Q_{T,i}^V(v)$. Transformation to the reference element $\hat{T}$ is again the main contribution. We transform, such that $\text{curl}\,\hat{v}(\hat{x}) = \text{curl}\,v(x), \text{curl}\,\hat{q}_i(\hat{x}) = \text{curl}\,q_i(x)$. Then, $\text{curl}\,v(\hat{x})\,\text{curl}\,q_i(\hat{x}) J_{[y_1,y_2,y_3]}$ lies in P^{2k} with respect to the y_i. The proof works the same way as the one above,

$$\begin{aligned}
Q_{T,i}^V(v) &= \int_{\omega_{V_1}} \int_{\omega_{V_2}} \int_{\omega_{V_3}} f_{V_1}(y_1) f_{V_2}(y_2) f_{V_3}(y_3) \int_{[y_1,y_2,y_3]} \operatorname{curl} v \cdot \operatorname{curl} q_i \, dx \, dy_3 dy_2 dy_1 \\
&= \int_{\hat{T}} \int_{\omega_{V_1}} f_{V_1}(y_1) \int_{\omega_{V_2}} f_{V_2}(y_2) \int_{\omega_{V_3}} f_{V_3}(y_3) \underbrace{\operatorname{curl} \hat{v}(\hat{x}) \cdot \operatorname{curl} \hat{q}_i(\hat{x}) J_{[y_1,y_2,y_3]}}_{\in P^{2k}} \, dy_3 dy_2 dy_1 \, d\hat{x} \\
&= \int_{\hat{T}} \operatorname{curl} \hat{v} \cdot \operatorname{curl} \hat{q}_i J_T \, d\hat{x} = \int_T \operatorname{curl} v \cdot \operatorname{curl} q_i \, dx.
\end{aligned}$$

□

6.2.1.2 Fundamental properties of $Q_{\alpha,k}^W, Q_{\alpha,k}^V$

We now concentrate on fundamental properties of the quasi-interpolation operators. Some of these were proven for the low-order case in [Sch01]. Extensions to the high-order case are lengthy and require much notation, but work quite straightforward; they can be done in a similar way as for the proof of Lemma 6.7. A key tool in our analysis will be the *commuting diagram property*, which means that interpolation and differentiation operators commute. In fact, for any $w \in H^1(\Omega_\alpha)$, $\alpha = x, z$, we have

$$Q_{\alpha,k}^V \nabla_\alpha w = \nabla_\alpha Q_{\alpha,k+1}^W w. \tag{6.29}$$

This property was shown for the low-order operators $Q_{\alpha,1}^W$ and $Q_{\alpha,0}^V$ in [Sch01], one verifies

$$N^V(Q_{\alpha,k}^V \nabla_\alpha w) = N^V(\nabla_\alpha Q_{\alpha,k+1}^W w),$$

for all degrees of freedom $N^V \in \mathcal{N}_{\alpha,k}^V$. This can be shown using basic calculus.

The following two lemmas state L^2 stability and optimal order approximation properties of the quasi-interpolation operators.

Lemma 6.8. *Let $\mathcal{T}_{h_\alpha}^\alpha$ be a shape-regular, quasi-uniform triangulation of Ω_α, $\alpha = x, z$. The quasi-interpolation operators $Q_{\alpha,k}^W, Q_{\alpha,k}^V$ are stable with respect to L^2, there exist constants c_1, c_2 independent of h_α such that*

$$\|Q_{\alpha,k}^W w\|_{L^2(\Omega_\alpha)} \leq c \|w\|_{L^2(\Omega_\alpha)}, \tag{6.30}$$
$$\|Q_{\alpha,k}^V v\|_{L^2(\Omega_\alpha)} \leq c \|v\|_{L^2(\Omega_\alpha)}. \tag{6.31}$$

Lemma 6.9. *On a shape-regular triangulation $\mathcal{T}_{h_\alpha}^\alpha$ of Ω_α, $\alpha = x, z$, let $Q_{\alpha,k}^W, Q_{\alpha,k}^V$ be the quasi-interpolation operators defined by relations (6.26), (6.27). For integers $l = 0, 1$ and $l < m \leq k+1$, let $w \in H^m(\Omega_\alpha)$ and $v \in [H^m(\Omega_\alpha)]^{\dim(\alpha)}$. Then, there exist constants $c_{QW}, c_{QV} > 0$ independent of*

h_α such that, for any element $T^\alpha \in \mathcal{T}_{h_\alpha}^\alpha$

$$\|w - \mathcal{Q}_{\alpha,k}^W w\|_{H^l(T^\alpha)} \leq c_{\mathcal{Q}^W} h_\alpha^{m-l} |w|_{H^m(\Delta_{T^\alpha})}, \tag{6.32}$$

$$\|v - \mathcal{Q}_{\alpha,k}^V v\|_{H^l(T^\alpha)} \leq c_{\mathcal{Q}^V} h_\alpha^{m-l} |v|_{H^m(\Delta_{T^\alpha})}. \tag{6.33}$$

One can show the statement of Lemma 6.8 using scaling arguments for the smoothed degrees of freedom, which are well-defined on L^2. The consistency of the operators with respect to polynomials up to order k ensures the approximation properties given in Lemma 6.9.

When defining the discrete norm $\|\cdot\|_{V_h}$, we used element-wise strains and normal-jumps across facets. We now show that these two quantities are approximated by the interpolation operator $\mathcal{Q}_{\alpha,k}^V$.

Lemma 6.10. *Let $\mathcal{T}_{h_\alpha}^\alpha$ be a shape-regular, quasi-uniform triangulation of Ω_α, $\alpha = x, z$. Let T^α be an arbitrary element, and $1 \leq m \leq k$. If v lies in $H^{m+1}(\Delta_{T^\alpha})$, the quasi-interpolation operator $\mathcal{Q}_{\alpha,k}^V$ for the Nédélec space satisfies the local error estimate*

$$\|\varepsilon_\alpha(v - \mathcal{Q}_{\alpha,k}^V v)\|_{T^\alpha} \leq c h_\alpha^m \|\nabla_\alpha^m \varepsilon_\alpha(v)\|_{\Delta_{T^\alpha}}. \tag{6.34}$$

The generic constant $c > 0$ only depends on the shape-regularity of $\mathcal{T}_{h_\alpha}^\alpha$, but not on its mesh size h_α.

Proof. Let α and T^α be fixed. In Lemma 6.7, it is shown that the quasi-interpolation operator $\mathcal{Q}_{\alpha,k}^V$ preserves polynomials up to order k on patches. Thus we may write, abbreviating $P^k = [P^k(\Delta_{T^\alpha})]^{\dim(\alpha)}$,

$$\begin{aligned}
\|\varepsilon_\alpha(v - \mathcal{Q}_{\alpha,k}^V v)\|_{T^\alpha} &= \inf_{q \in P^k} \|\varepsilon_\alpha\big((id - \mathcal{Q}_{\alpha,k}^V)(v - q)\big)\|_{T^\alpha} \\
&\leq \inf_{q \in P^k} \Big[\|\varepsilon_\alpha(v - q)\|_{T^\alpha} + \|\varepsilon_\alpha\big(\mathcal{Q}_{\alpha,k}^V(v - q)\big)\|_{T^\alpha}\Big].
\end{aligned}$$

Next, we apply an inverse inequality on the finite-dimensional local space V_α^k. To do so, we have to verify, that Korn's inequality holds for both $v - q$ and $\mathcal{Q}_{\alpha,k}^V(v - q)$ on T^α. As the interpolation operator preserves rigid body motions, and $RM(\Delta_{T^\alpha})$ is contained in $[P^k(\Delta_{T^\alpha})]^{\dim(\alpha)}$, this is true, and we may estimate

$$\begin{aligned}
\|\varepsilon_\alpha(v - \mathcal{Q}_{\alpha,k}^V v)\|_{T^\alpha} &\leq c h_\alpha^{-1} \inf_{q \in P^k} \Big[\|v - q\|_{T^\alpha} + \|\mathcal{Q}_{\alpha,k}^V(v - q)\|_{T^\alpha}\Big] \\
&\leq c h_\alpha^{-1} \inf_{q \in P^k} \|v - q\|_{T^\alpha}.
\end{aligned}$$

In the last line, we employed the L^2 stability of $\mathcal{Q}_{\alpha,k}^V$ (Lemma 6.8). Next, we apply the Lemma of Bramble and Hilbert, and see that, for $0 \leq m \leq k$

$$\|\varepsilon_\alpha(v - \mathcal{Q}_{\alpha,k}^V v)\|_{T^\alpha} \leq c h_\alpha^m \|\nabla_\alpha^{m+1} v\|_{\Delta_{T^\alpha}}.$$

For $\alpha = x$, one can show $\|\nabla_x^m \varepsilon_x(v)\|_{\Delta_{T^x}} \simeq \|\nabla_x^{m+1} v\|_{\Delta_{T^x}}$ for $m \geq 1$ by a direct evaluation of the respective terms. For $\alpha = z$, the strain and gradient operator coincide. Putting all estimates together, we obtain
$$\|\varepsilon_\alpha(v - \mathcal{Q}_{\alpha,k}^V v)\|_{T^\alpha} \leq c h_\alpha^m \|\nabla_\alpha^m \varepsilon_\alpha(v)\|_{\Delta_{T^\alpha}},$$
which concludes the proof of the lemma. $\square$

Lemma 6.11. *Let $\mathcal{T}_{h_\alpha}^\alpha$ be a shape-regular, quasi-uniform triangulation of Ω_α, $\alpha = x, z$. Let F^α be an arbitrary facet with neighbors T_1^α, T_2^α, and $1 \leq m \leq k$. If v lies in $H^{m+1}(\Omega_\alpha)$, the quasi-interpolation operator $\mathcal{Q}_{\alpha,k}^V$ for the Nédélec space satisfies the error estimate*
$$\|[v - \mathcal{Q}_{\alpha,k}^V v]\|_{F^\alpha} \leq c h_\alpha^{m+1/2} \sum_{i=1}^2 \|\nabla_\alpha^m \varepsilon_\alpha(v)\|_{\Delta_{T_i^\alpha}}. \tag{6.35}$$

The generic constant $c > 0$ only depends on the shape-regularity of $\mathcal{T}_{h_\alpha}^\alpha$, but not on its mesh size h_α.

Proof. In Lemma 4.27, we proved a Korn-type inequality of the form
$$\|\nabla_\alpha v_\alpha\|_{T^\alpha}^2 \leq c \left(\|\varepsilon_\alpha(v_\alpha)\|_{T^\alpha}^2 + h_\alpha^{-2} \|\mathcal{I}_{\alpha,0}^V v_\alpha\|_{T^\alpha}^2 \right)$$
(cf. equation (4.36)), for an element T^α in the shape-regular triangulation $\mathcal{T}_{h_\alpha}^\alpha$. Recalling the proof, one can see that a similar estimate holds, where the nodal interpolation operator $\mathcal{I}_{\alpha,0}^V$ is replaced by $\mathcal{Q}_{\alpha,0}^V$:
$$\|\nabla_\alpha v_\alpha\|_{T^\alpha}^2 \leq c \left(\|\varepsilon_\alpha(v_\alpha)\|_{T^\alpha}^2 + h_\alpha^{-2} \|\mathcal{Q}_{\alpha,0}^V v_\alpha\|_{T^\alpha}^2 \right). \tag{6.36}$$

Using this inequality and a scaled trace inequality we proceed
$$\begin{aligned}
h_\alpha^{-1} \|[(id - \mathcal{Q}_{\alpha,k}^V) v]\|_{F^\alpha}^2 &\leq \sum_{i=1}^2 \|\nabla_\alpha(v - \mathcal{Q}_{\alpha,k}^V v)\|_{T_i^\alpha}^2 \\
&\leq \sum_{i=1}^2 \left[\|\varepsilon_\alpha(v - \mathcal{Q}_{\alpha,k}^V v)\|_{T_i^\alpha}^2 + h_\alpha^{-2} \|\mathcal{Q}_{0,h}^V(v_\alpha - \mathcal{Q}_{\alpha,k}^V v_\alpha)\|_{T_i^\alpha}^2 \right] \\
&= \sum_{i=1}^2 \|\varepsilon_\alpha(v - \mathcal{Q}_{\alpha,k}^V v)\|_{T_i^\alpha}^2.
\end{aligned}$$

Note that the last equality can be derived from the hierarchical definition of the quasi-interpolation operator, which ensures that $\mathcal{Q}_{\alpha,0}^V = \mathcal{Q}_{\alpha,0}^V \mathcal{Q}_{\alpha,k}^V$. Now, the approximation property of the strain (see Lemma 6.10) implies
$$h_\alpha^{-1} \|[(id - \mathcal{Q}_{\alpha,k}^V) v]\|_{F^\alpha}^2 \leq c h_\alpha^{2m} \sum_{i=1}^2 \|\nabla_\alpha^m \varepsilon_\alpha(v)\|_{\Delta_{T_i^\alpha}}^2,$$
which completes the proof. $\square$

6.2.2 Tensor product interpolation

In this section, we derive interpolation operators $\mathcal{Q}_{h,k}^V$ and $\mathcal{I}_{h,k}^\Sigma$ for the spaces V_h^k and Σ_h^k on the tensor product mesh $\mathcal{T}_h$. We propose to use tensor products of the respective operators in one and two dimensions, as the finite element spaces are of this structure. In order to define them, we assume that all functions are square integrable with respect to x and z separately, and that Fubini's theorem is applicable. Then, we define the quasi-interpolation operator for the displacement space by

$$\mathcal{Q}_{h,k}^V(v) := \begin{pmatrix} \mathcal{Q}_{x,k}^V \otimes \mathcal{Q}_{z,k+1}^W v_x \\ \mathcal{Q}_{x,k+1}^W \otimes \mathcal{Q}_{z,k}^V v_z \end{pmatrix} \quad \text{for } v = (v_x, v_z)^T \in [L^2(\Omega)]^d. \tag{6.37}$$

Thus, the in-plane deformation v_x is interpolated by the quasi-interpolation operator for the Nédélec space in the plane, and continuously in thickness direction. For the transversal displacement v_z, the setting is vice versa: in-plane, it is interpolated continuously, whereas we perform an L^2 projection in thickness direction. For the stress space in three dimensions, we define

$$\mathcal{I}_{h,k}^\Sigma(\tau) := \begin{pmatrix} (\mathcal{I}_{x,k}^\Sigma \otimes \mathcal{I}_{z,k+1}^P)\tau_x & (\mathcal{I}_{x,k}^P \otimes \mathcal{I}_{z,k+1}^P)\tau_{xz} \\ \text{sym} & (\mathcal{I}_{x,k+1}^P \otimes \mathcal{I}_{z,k+1}^\Sigma)\tau_z \end{pmatrix} \quad \text{for } \tau \in H(\text{div div}; \Omega). \tag{6.38}$$

For planar problems, i.e. $d = 2$, one needs to modify the order of the upper-left operator to $\mathcal{I}_{x,k+1}^\Sigma \otimes \mathcal{I}_{z,k+1}^P$, such that it matches the finite element space.

Our goal is to obtain anisotropic interpolation error estimates

$$\|v - \mathcal{Q}_{h,k}^V v\|_{V_h} \leq c\bigl(h_x^m \|\nabla_x^m \varepsilon(v)\|_\Omega + h_z^m \|\nabla_z^m \varepsilon(v)\|_\Omega\bigr), \tag{6.39}$$

$$\|\tau - \mathcal{I}_{h,k}^\Sigma \tau\|_{\Sigma_h} \leq c\bigl(h_x^{m+1} \|\nabla_x^{m+1} \tau\|_\Omega + h_z^{m+1} \|\nabla_z^{m+1} \tau\|_\Omega\bigr) \tag{6.40}$$

for $m \leq k$ and τ, v m times weakly differentiable. Note that we have an additional power of h_x, h_z for the stresses, as the differential order of the stress norm is one less than for the displacement norm.

We split the proof for the first estimate (6.39) into two parts. As the norm $\|\cdot\|_{V_h}$ consists of contributions from the piecewise defined strain operator, as well as jump terms, we do these estimates separately. First, we derive an interpolation error estimate for the strain. Then, we concentrate on the jump of the normal displacements. In both cases, we obtain the desired estimates, which then imply the bound (6.39).

Lemma 6.12. *Let $T = T^x \times T^z$ be an element of $\mathcal{T}_h$. Let, for integer $m \leq k$, be $v \in [H^{m+1}(\Delta_T)]^d$. Then the quasi-interpolation operator $\mathcal{Q}_{h,k}^V$ according to its definition (6.37) satisfies the local anisotropic error estimate*

$$\|\varepsilon(v - \mathcal{Q}_{h,k}^V v)\|_T \leq c\bigl(h_x^m \|\nabla_x^m \varepsilon(v)\|_{\Delta_T} + h_z^m \|\nabla_z^m \varepsilon(v)\|_{\Delta_T}\bigr). \tag{6.41}$$

Proof. Let $v \in H^{m+1}(\Delta_T)$ be fixed. We bound the different blocks $\varepsilon_x(v)$, $\varepsilon_z(v)$, and $\varepsilon_{xz}(v)$

separately. For the diagonal blocks $\varepsilon_x(v)$, $\varepsilon_z(v)$, the estimates are obtained in a similar manner, therefore we treat them together. We choose $\alpha, \beta = x, z$, with $\alpha \neq \beta$. We note that the diagonal sub-blocks of the strain depend only on the respective component of the displacement, and so we may write $\varepsilon_\alpha(v_\alpha)$ and $\varepsilon_\beta(v_\beta)$. We use that interpolation and differentiation with respect to different variables commute, i.e.

$$\mathcal{Q}^V_{\alpha,k} \nabla_\beta v_\beta = \nabla_\beta \mathcal{Q}^V_{\alpha,k} v_\beta.$$

We estimate, using moreover boundedness in L^2 (Lemma 6.8) and approximation properties (Lemma 6.9) of $\mathcal{Q}^L_{\beta,k+1}$, and the approximation of the strain by $\mathcal{Q}^V_{\alpha,k}$ (Lemma 6.10),

$$\begin{aligned}
\frac{1}{2}\|\varepsilon_\alpha(v - \mathcal{Q}^V_{h,k}v)\|^2_T &= \frac{1}{2}\|\varepsilon_\alpha(v_\alpha - \mathcal{Q}^V_{\alpha,k}\mathcal{Q}^W_{\beta,k+1}v_\alpha)\|^2_T \\
&\leq \|(id - \mathcal{Q}^W_{\beta,k+1})\varepsilon_\alpha(v_\alpha)\|^2_T + \|\mathcal{Q}^W_{\beta,k+1}\varepsilon_\alpha(v_\alpha - \mathcal{Q}^V_{\alpha,k}v_\alpha)\|^2_T \\
&\leq c\left(h_\beta^{2m}\|\nabla_\beta^m \varepsilon_\alpha(v_\alpha)\|^2_{\Delta_T} + \|\varepsilon_\alpha(v - \mathcal{Q}^V_{\alpha,k}v_\alpha)\|^2_T\right) \\
&\leq c\left(h_\beta^{2m}\|\nabla_\beta^m \varepsilon_\alpha(v_\alpha)\|^2_{\Delta_T} + h_\alpha^{2m}\|\nabla_\alpha^m \varepsilon_\alpha(v_\alpha)\|^2_{\Delta_T}\right).
\end{aligned}$$

Next, we concentrate on the off-diagonal block $\varepsilon_{xz}(v)$. In the following calculation, we use the commuting diagram property of the quasi-interpolation operators (6.29), and again their L^2 continuity and approximation properties.

$$\begin{aligned}
\frac{1}{2}\|\varepsilon_{xz}(v - \mathcal{Q}^V_{h,k}v)\|^2_T &= \frac{1}{8}\|\nabla_x(v_z - \mathcal{Q}^V_{z,k}\mathcal{Q}^W_{x,k+1}v_z) + \nabla_z(v_x - \mathcal{Q}^W_{z,k+1}\mathcal{Q}^V_{x,k}v_x)\|^2_T \\
&= \frac{1}{8}\|(id - \mathcal{Q}^V_{z,k}\mathcal{Q}^V_{x,k})\underbrace{(\nabla_x v_x + \nabla_z v_z)}_{2\varepsilon_{xz}(v)}\|^2_T \\
&\leq \|(id - \mathcal{Q}^V_{x,k})\varepsilon_{xz}(v)\|^2_T + \|\mathcal{Q}^V_{x,k}(id - \mathcal{Q}^V_{z,k})\varepsilon_{xz}(v)\|^2_T \\
&\leq c\left(h_x^{2m}\|\nabla_x^m \varepsilon_{xz}(v)\|^2_{\Delta_T} + h_z^{2m}\|\nabla_z^m \varepsilon_{xz}(v)\|^2_{\Delta_T}\right).
\end{aligned}$$

Thereby, we conclude the proof. $\square$

Lemma 6.13. *Let $\alpha = x, z$, and let $F \in \mathcal{F}_h$ be a facet with normal n in direction α, i.e. an in-plane facet for $\alpha = z$ and a transversal one for $\alpha = x$. For $m \leq k$, and $v \in [H^{m+1}(\Omega)]^d$, the jump of the normal component $[\![v]\!]_n$ across F is approximated by $\mathcal{Q}^V_{h,k}$ at an optimal order, i.e.*

$$\|[\![v - \mathcal{Q}^V_{h,k}v]\!]_n\|_F \leq c \sum_{T \in \Delta_F} h_\alpha^{m+1/2} \|\nabla_\alpha^m \varepsilon_\alpha(v)\|_{\Delta_T}. \tag{6.42}$$

Proof. Let $\beta = x, z$ denote the direction orthogonal to α, i.e. $\beta \neq \alpha$. As F is orthogonal to α, we have that $h_F = h_\alpha$, and that the normal jump $[\![w]\!]_{n,F}$ of a piecewise smooth function w depends only on the component w_α. Thus, we have, for any such function

$$\|[w]_n\|_F \leq \|[w_\alpha]\|_F.$$

We apply this to $v - \mathcal{Q}^V_{h,k}v$, and obtain, using the definition of the quasi-interpolation operator and the triangle inequality,

$$\begin{aligned}
\|[v - \mathcal{Q}^V_{h,k}v]_n\|_F &\leq \|[v_\alpha - \mathcal{Q}^W_{\beta,k+1}\mathcal{Q}^V_{\alpha,k}v_\alpha]\|_F \\
&\leq \|[(id - \mathcal{Q}^W_{\beta,k+1})v]\|_F + \|[\mathcal{Q}^W_{\beta,k+1}(id - \mathcal{Q}^V_{\alpha,k})v_\alpha]\|_F^2.
\end{aligned}$$

We treat the two terms separately. We first show that the first term vanishes, as v_α is continuous across F. Let therefore T_1, T_2 be the elements neighboring F, then

$$\begin{aligned}
\|[(id - \mathcal{Q}^W_{\beta,k+1})v_\alpha]\|_F &= \|(v_\alpha - \mathcal{Q}^W_{\beta,k+1}v_\alpha)|_{T_1} - (v_\alpha - \mathcal{Q}^W_{\beta,k+1}v_\alpha)|_{T_2}\|_F \\
&= \|(v_\alpha|_{T_1} - v_\alpha|_{T_2}) - \mathcal{Q}^W_{\beta,k+1}(v_\alpha|_{T_1} - v_\alpha|_{T_2})\|_F = 0
\end{aligned}$$

We concentrate on the second term in the estimate above. Boundedness of $\mathcal{Q}^W_{\beta,k+1}$ in L^2 implies

$$\|[\mathcal{Q}^W_{\beta,k+1}(id - \mathcal{Q}^V_{\alpha,k})v_\alpha]\|_F^2 \leq c\|[(id - \mathcal{Q}^V_{\alpha,k})v_\alpha]\|_F^2.$$

Now, we note that the facet F is the tensor product of a simplicial element T^β and a facet F^α. We assume v to be sufficiently smooth such that Fubini's theorem is applicable. Lemma 6.11 bounds the interpolation error of the jump across facet F^α in terms of the α block of the strain. We use this estimate and obtain

$$\begin{aligned}
\|[(id - \mathcal{Q}^V_{\alpha,k})v_\alpha]\|_F^2 &= \int_{T^\beta} \int_{F^\alpha} [(id - \mathcal{Q}^V_{\alpha,k})v_\alpha]\, ds_\alpha d\beta \\
&\leq h^{2m+1} \sum_{T^\alpha \in \Delta_{F^\alpha}} \int_{T^\beta} \int_{T^\alpha} \nabla^m_\alpha \varepsilon_\alpha(v)\, d\alpha\, d\beta \\
&= h^{2m+1} \sum_{T \in \Delta_F} \|\nabla^m_\alpha \varepsilon_\alpha(v)\|^2_T.
\end{aligned}$$

$\square$

Together, Lemmas 6.12 and 6.13 imply an interpolation error estimate with respect to the discrete norm $\|\cdot\|_{V_h}$.

Theorem 6.14. *Let Ω be a tensor product domain according to Definition 6.1, and $\mathcal{T}_h$ the corresponding triangulation. Then, for integers $1 \leq m \leq k$, and $v \in H^m(\Omega)$, the interpolation error can be bounded by*

$$\|v - \mathcal{Q}^V_{h,k}v\|_{V_h} \leq c\Big(h^m_{\boldsymbol{x}}\|\nabla^m_{\boldsymbol{x}}\varepsilon(v)\|_\Omega + h^m_z\|\nabla^m_z\varepsilon(v)\|_\Omega\Big). \tag{6.43}$$

The generic constant $c > 0$ does not depend on the mesh sizes $h_{\boldsymbol{x}}, h_z$ or their ratio $h_{\boldsymbol{x}}/h_z$.

Thus, we have shown the desired inequality (6.39), which gives an optimal order of convergence for the interpolation error with respect to the displacement norm $\|\cdot\|_{V_h}$. We now concentrate on verifying estimate (6.40), which concerns the approximation properties of the stress space.

Theorem 6.15. *Let Ω be a tensor product domain as in Definition 6.1, with corresponding triangulation $\mathcal{T}_h$. Let $1 \leq m \leq k+1$, and $\tau \in H^m_{SYM}(\Omega)$. The interpolation operator $\mathcal{I}^{\Sigma}_{h,k}$ satisfies the approximation property*

$$\|\tau - \mathcal{I}^{\Sigma}_{h,k}\tau\|_{\Sigma_h} \leq c\left(h^m_{\boldsymbol{x}}\|\nabla^m_{\boldsymbol{x}}\tau\|_{\Omega} + h^m_z\|\nabla^m_z\tau\|_{\Omega}\right), \qquad (6.44)$$

where c is independent of $h_{\boldsymbol{x}}, h_z$ and $h_{\boldsymbol{x}}/h_z$.

Proof. We bound the different sub-blocks $\tau_{\boldsymbol{x}}, \tau_z$ and $\tau_{\boldsymbol{x}z}$. As the three estimates are obtained in a similar way, we do the calculations for the block $\tau_{\boldsymbol{x}}$, and $d = 3$. The tensor product definition and the triangle inequality yield

$$\begin{aligned}
\|\tau_{\boldsymbol{x}} - \mathcal{I}^{\Sigma}_{h,k}\tau_{\boldsymbol{x}}\|_{\Sigma_h} &= \|\tau_{\boldsymbol{x}} - \mathcal{I}^{\Sigma}_{\boldsymbol{x},k}\mathcal{I}^{P}_{z,k+1}\tau_{\boldsymbol{x}}\|_{\Sigma_h} \\
&\leq \|\tau_{\boldsymbol{x}} - \mathcal{I}^{P}_{z,k+1}\tau_{\boldsymbol{x}}\|_{\Sigma_h} + \|\mathcal{I}^{P}_{z,k+1}(\tau_{\boldsymbol{x}} - \mathcal{I}^{\Sigma}_{\boldsymbol{x},k}\tau_{\boldsymbol{x}})\|_{\Sigma_h}.
\end{aligned}$$

Next, we use the L^2 continuity of the projection operator $\mathcal{I}^{P}_{z,k+1}$ and approximation properties of both operators, and obtain

$$\|\tau_{\boldsymbol{x}} - \mathcal{I}^{\Sigma}_{h,k}\tau_{\boldsymbol{x}}\|_{\Sigma_h} \leq c\left(h^m_z\|\nabla^m_z\tau_{\boldsymbol{x}}\|_{\Omega} + h^m_{\boldsymbol{x}}\|\nabla^m_{\boldsymbol{x}}\tau_{\boldsymbol{x}}\|_{\Omega}\right).$$

This estimate, together with the respective ones for $\tau_z, \tau_{\boldsymbol{x}z}$, completes the proof. □

6.2.3 Anisotropic error estimates

We will now collect all results on the TD-NNS method in anisotropic domains, and provide an a-priori error estimate for the finite element solution (σ_h, u_h).

Theorem 6.16. *Let Ω be a tensor product domain as in Definition 6.1, with matching triangulation $\mathcal{T}_h$. Let k, m be positive integers such that $m \leq k$. Let $(\sigma, u) \in H^m_{SYM}(\Omega) \times [H^{m+1}(\Omega)]^d$ be the solution to the elasticity problem (Problem 3.1). Then, the finite element solution $(\sigma_h, u_h) \in \Sigma^k_h \times V^k_h$ to Problem 6.2 satisfies the a-priori error bound*

$$\begin{aligned}
\|\sigma - \sigma_h\|_{\Sigma_h} + \|u - u_h\|_{V_h} &\leq c\Big(h^m_{\boldsymbol{x}}\|\nabla^m_{\boldsymbol{x}}\tau\|_{\Omega} + h^m_z\|\nabla^m_z\tau\|_{\Omega} + \\
&\qquad h^m_{\boldsymbol{x}}\|\nabla^m_{\boldsymbol{x}}\varepsilon(v)\|_{\Omega} + h^m_z\|\nabla^m_z\varepsilon(v)\|_{\Omega}\Big).
\end{aligned}$$

The constant $c > 0$ is independent of $h_{\boldsymbol{x}}, h_z$ and $h_{\boldsymbol{x}}/h_z$.

Proof. The theorem is a consequence of the basic error estimate for mixed problems, see Lemma 4.3. We use that all conditions of Brezzi's theorem (Assumption 4.2) are satisfied in the discrete setting,

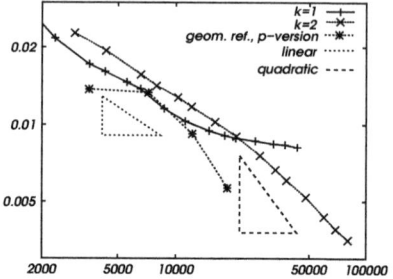

Figure 6.2: Plate, aspect ratio 1:100, methods of order $k = 1, 2$ and p-FEM. We plot coupling degrees of freedom vs. estimated error $\|\sigma - \sigma_h\|_{\Sigma_h}$.

and the interpolation error estimates from the previous section. As we ensured in all manipulations, the constant is independent as well of the anisotropic mesh sizes, as of their aspect ratio. □

6.3 Numerical results

We present some numerical results obtained using the tensor product elements described above. We first consider a plate of unit size, where the thickness is 0.01. We assume that the plate is clamped on one side, and pulled horizontally on the opposite boundary. We discretize it in-plane with a triangular mesh consisting of 46 elements. We use one layer of elements in thickness direction. We do computations using methods of order $k = 1, 2$. In Figure 6.2, we plot the results obtained when doing adaptive refinement in x direction. For the error estimation, we use a Zienkiewicz-Zhu type estimator [ZZ87]. We see that the convergence is not completely of optimal order, which is due to the constant mesh size h_z in vertical direction. Moreover, we use the hp-version of the finite element method. Therefore, we do one level of geometric refinement towards the sides of the plate, and then increase the polynomial order of the method. We see the expected exponential convergence in Figure 6.2. In Figure 6.3, we display the stress component $\sigma_{h,x_1 x_2}$ obtained for the method of order 4, and zoom into the vicinity of a corner, where a singularity arises.

As a second example, we do computations on a plate with hole, which is of thickness $d_z = 0.005$. We assume that the plate is clamped along one side, and a constant surface traction in normal direction is acting on the opposite direction. We use two levels of geometric refinement In Figure 6.4, we plot the absolute value of the stress. We use a finite element method of order $k = 3$.

Next, we apply the TD-NNS method to non-tensor product domains. We consider a cylindrical shell, which is fixed along three sides, and where a volume force $f = (0, 0, xyz)$ is acting in vertical direction. The thickness of the shell is set to $d_z = 0.005$. We use both the pure displacement formulation with standard high-order H^1 elements, and our method. In Figure 6.5, we plot the absolute value of the stress $|\sigma_h|$, which we obtain for the two methods, where we choose $k = 3$ for

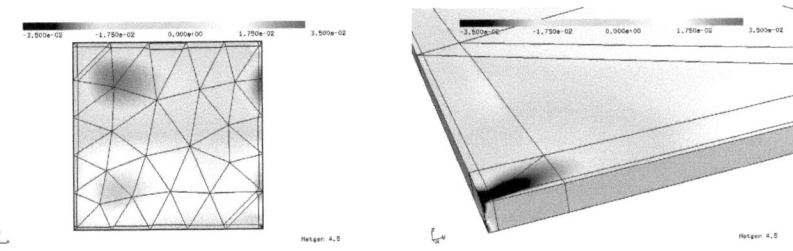

Figure 6.3: Plate, aspect ratio 1:100, method of order $k = 4$. We plot the stress component σ_{h,x_1x_2}, and zoom into the vicinity of a singularity.

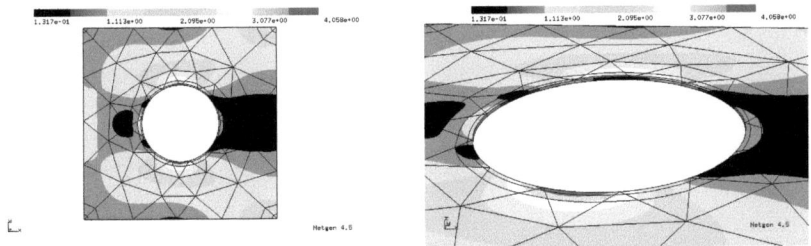

Figure 6.4: Plate with hole, aspect ratio 1:200, two levels of geometric refinement. We plot the absolute value of the stress $|\sigma_h|$.

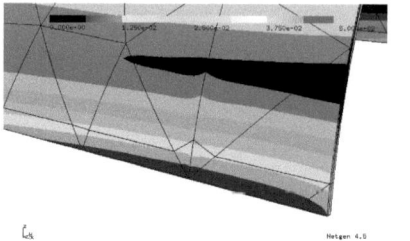 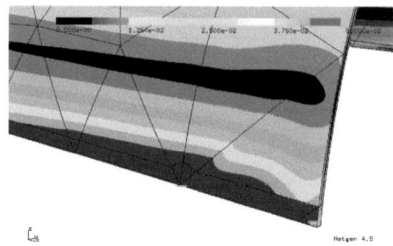

Figure 6.5: Cylindrical shell, aspect ratio 1:200, one level of geometric refinement. We plot $|\sigma_h|$. Left: TD-NNS method for $k = 3$, 46 949 coupling dofs. Right: standard H^1 method for $k = 5$, 54 918 coupling dofs.

the TD-NNS elements, and order $k = 5$ for the standard method. This results in 46 949 globally coupling dofs for the TD-NNS, and 54 918 dofs for the standard method. We see the far better quality of the TD-NNS solution, as it does not suffer from shear locking. Again, we use geometric mesh refinement to resolve the singularities expected at the corners of the shell. The two lower plots show the computed solution close to the corner point.

In our last example, we couple the TD-NNS method for anisotropic domains to the standard TD-NNS method on a shape-regular mesh. We consider a geometry consisting of a cylinder, which is partly coated by a different material. The inner cylinder is of unit size, while the coating is of thickness $d_z = 0.01$. We assume that the inner cylinder is characterized by a Young's modulus $\bar{E} = 1$, while its Poisson ratio is set to $\bar{\nu} = 0.4999$. For the coating, we choose $\bar{E} = 1\,000, \bar{\nu} = 0.4$. The body is fixed on the left end, and a volume force in vertical direction is acting. We discretize the inner domain by a shape-regular, tetrahedral mesh, and the coating by a prismatic one. In total, we use 924 elements. Then, we apply the hybridized TD-NNS method from Chapter 5, where we use the tensor product elements derived above for the prismatic elements. For a method of order $k = 3$, we obtain 209 849 degrees of freedom, of which 55 457 are coupling. We apply a preconditioned CG method, which needs 174 steps for an error reduction of 10^{-8}. Here, we used the additive Schwarz block preconditioner from Chapter 5, enhanced by blocks containing all degrees of freedom of a flat prism, and those of the neighboring tetrahedral element. This was done to avoid a slow-down of convergence due to the anisotropic nature of the prismatic elements. In Figure 6.6, we display geometry, mesh, and stresses in the deformed geometry.

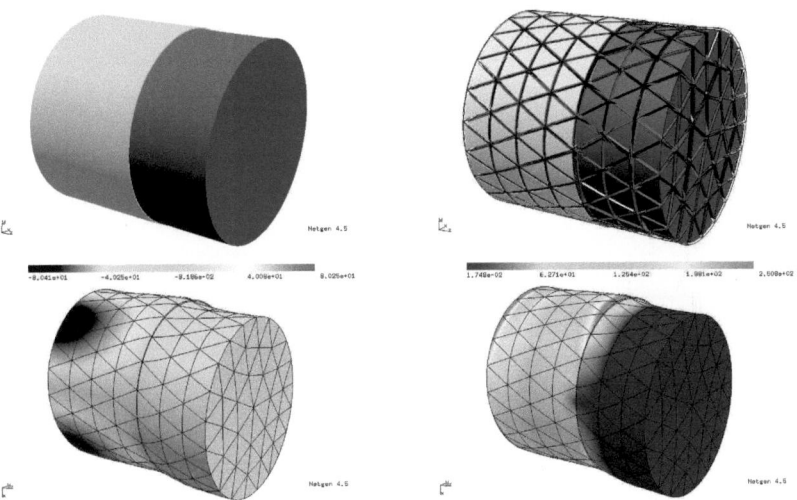

Figure 6.6: Coated cylinder, upper left: geometry, upper right: hybrid mesh, lower left: stress component $\sigma_{h,yz}$, lower right: absolute value of stress $|\sigma_h|$ on deformed configuration for method of order $k = 3$.

Bibliography

[AAFM99] Stephen M. Alessandrini, Douglas N. Arnold, Richard S. Falk, and Alexandre L. Madureira. Derivation and justification of plate models by variational methods. In *Plates and shells (Québec, QC, 1996)*, volume 21 of *CRM Proc. Lecture Notes*, pages 1–20. Amer. Math. Soc., Providence, RI, 1999.

[AAW08] D. N. Arnold, G. Awanou, and R. Winther. Finite elements for symmetric tensors in three dimensions. *Math. Comp.*, 77(263):1229–1251, 2008.

[AB85] D. N. Arnold and F. Brezzi. Mixed and nonconforming finite element methods: implementation, postprocessing and error estimates. *RAIRO Math. Mod. Anal. Numer.*, 19:7–32, 1985.

[ABCM02] D. N. Arnold, F. Brezzi, B. Cockburn, and L.D. Marini. Unified analysis of discontinuous Galerkin methods for elliptic problems. *SIAM J. Numer. Anal.*, 39(5):1749–1779, 2002.

[ABD84] D. N. Arnold, F. Brezzi, and J. Douglas. PEERS: a new finite element for plane elasticity. *Jap. J. Appl. Math.*, 1:347–367, 1984.

[AC03] M. Ainsworth and J. Coyle. Hierarchic finite element bases on unstructured tetrahedral meshes. *Internat. J. Numer. Methods Engrg.*, 58(14):2103–2130, 2003.

[AC05] S. Adams and B. Cockburn. A mixed finite element method for elasticity in three dimensions. *J. Sci. Comput.*, pages 515–521, 2005.

[ADG84] D. N. Arnold, J. Douglas, and C. P. Gupta. A family of higher order mixd finite element methods for plane elasticity. *Numer. Math.*, 45:1–22, 1984.

[AFW97a] D. N. Arnold, R. S. Falk, and R. Winther. Preconditioning discrete approximations of the Reissner-Mindlin plate model. *RAIRO Modél. Math. Anal. Numér.*, 31(4):517–557, 1997.

[AFW97b] D. N. Arnold, R. S. Falk, and R. Winther. Preconditioning in $H(\text{div})$ and applications. *Math. Comp.*, 66(219):957–984, 1997.

[AFW98]   D. N. Arnold, R. S. Falk, and R. Winther. Multigrid preconditioning in $H(\text{div})$ on non-convex polygons. *Comput. Appl. Math.*, 17(3):303–315, 1998.

[AFW00]   D. N. Arnold, R. Falk, and R. Winther. Multigrid in $H(\text{div})$ and $H(\text{curl})$. *Numer. Math.*, 85(2):197–217, 2000.

[AFW07]   D. N. Arnold, R. S. Falk, and R. Winther. Mixed finite element methods for linear elasticity with weakly imposed symmetry. *Math. Comp.*, 76(260):1699–1723 (electronic), 2007.

[Ain96a]   M. Ainsworth. A hierarchical domain decomposition preconditioner for h-p finite element approximation on locally refined meshes. *SIAM J. Sci. Comput.*, 17(6):1395–1413, 1996.

[Ain96b]   M. Ainsworth. A preconditioner based on domain decomposition for h-p finite-element approximation on quasi-uniform meshes. *SIAM J. Numer. Anal.*, 33(4):1358–1376, 1996.

[Ain98]   M. Ainsworth. A posteriori estimation for fully discrete hierarchical models of elliptic boundary value problems on thin domains. *Numer. Math.*, 80:325–362, 1998.

[AW02]   D. N. Arnold and R. Winther. Mixed finite elements for elaticity. *Numer. Math*, 92:401–419, 2002.

[BC01a]   A. Buffa and P. Ciarlet, Jr. On traces for functional spaces related to Maxwell's equations Part I: An integration by parts formula in Lipschitz polyhedra. *Math. Methods Appl. Sci*, 24(1):9–30, 2001.

[BC01b]   A. Buffa and P. Ciarlet, Jr. On traces for functional spaces related to Maxwell's equations Part II: Hodge decompositions on the boundary of Lipschitz polyhedra and applications. *Math. Methods Appl. Sci*, 24(1):31–48, 2001.

[BCMS06]   F. Brezzi, B. Cockburn, L. D. Marini, and E. Süli. Stabilization mechanisms in discontinuous Galerkin finite element methods. *Comput. Methods Appl. Mech. Engrg.*, 195(25-28):3293–3310, 2006.

[BD81]   I. Babuška and M. R. Dorr. Error estimates for the combined h and p versions of the finite element method. *Numer. Math.*, 37(2):257–277, 1981.

[BDM85]   F. Brezzi, J. Douglas, and L.D. Marini. Two families of mixed finite elements for second order elliptic problems. *Numer. Math.*, 24:217–235, 1985.

[BF91]   F. Brezzi and M. Fortin. *Mixed and Hybrid Finite Element Methods*. Springer-Verlag, New-York, 1991.

[BL91]   I. Babuška and L. Li. Hierarchic modeling of plates. *Computers and Structures*, 40(2):419–430, 1991.

[BPX90] J. H. Bramble, J. E. Pasciak, and J. Xu. Parallel multilevel preconditioners. *Math. Comp.*, 55(191):1–22, 1990.

[Bra92] D. Braess. *Finite Elemente: Theorie, schnelle Löser und Anwendungen in der Elastizitätstheorie*. Springer Verlag, Berlin, 1992.

[Bra93] J. H. Bramble. *Multigrid methods*, volume 294 of *Pitman Research Notes in Mathematics Series*. Longman Scientific & Technical, Harlow, 1993.

[Bre74] F. Brezzi. On the existence, uniqueness and approximation of saddle-point problems arising from Lagrangian multipliers. *RAIRO*, 8:129–151, 1974.

[Bre96] S. C. Brenner. Multigrid methods for parameter dependent problems. *RAIRO Modél. Math. Anal. Numér.*, 30(3):265–297, 1996.

[Bre04] S. C. Brenner. Korn's inequalities for piecewise H^1 vector fields. *Mathematics of Computation*, 73:1067–1087, 2004.

[BS02]   S. C. Brenner and L. R. Scott. *The Mathematical Theory of Finite Element Methods*. Springer-Verlag, New York, 2002.

[BSK81] I. Babuška, B. A. Szabo, and I. N. Katz. The p-version of the finite element method. *SIAM J. Numer. Anal.*, 18(3):515–545, 1981.

[BZ00]   J. H. Bramble and X. Zhang. The analysis of multigrid methods. In *Handbook of numerical analysis, Vol. VII*, Handb. Numer. Anal., VII, pages 173–415. North-Holland, Amsterdam, 2000.

[CCS06] J. Carrero, B. Cockburn, and D. Schötzau. Hybridized globally divergence-free LDG methods. I. The Stokes problem. *Math. Comp.*, 75(254):533–563 (electronic), 2006.

[CG04]   B. Cockburn and J. Gopalakrishnan. A characterization of hybridized mixed methods for second order elliptic problems. *SIAM J. Numer. Anal.*, 42(1):283–301 (electronic), 2004.

[CG05a] B. Cockburn and J. Gopalakrishnan. Error analysis of variable degree mixed methods for elliptic problems via hybridization. *Math. Comp.*, 74(252):1653–1677 (electronic), 2005.

[CG05b] B. Cockburn and J. Gopalakrishnan. Incompressible finite elements via hybridization. I. The Stokes system in two space dimensions. *SIAM J. Numer. Anal.*, 43(4):1627–1650 (electronic), 2005.

[CG05c]   B. Cockburn and J. Gopalakrishnan. Incompressible finite elements via hybridization. II. The Stokes system in three space dimensions. *SIAM J. Numer. Anal.*, 43(4):1651–1672 (electronic), 2005.

[CGL07]   B. Cockburn, J. Gopalakrishnan, and R. Lazarov. Unified hybridization of discontinuous Galerkin, mixed and conforming Galerkin methods for second order elliptic problems. *preprint*, 2007.

[Cia78]   P. G. Ciarlet. *The Finite Element Method for Elliptic Problems. Studies in Mathematics and Applications*. North-Holland Publishing Co., Amsterdam, New York, Oxford, 1978.

[Cia88]   P. G. Ciarlet. *Mathematical elasticity. Vol. I*, volume 20 of *Studies in Mathematics and its Applications*. North-Holland Publishing Co., Amsterdam, 1988. Three-dimensional elasticity.

[Cia91]   P. G. Ciarlet. Basic error estimates for elliptic problems. In P. G. Ciarlet and J.-L. Lions, editors, *Handbook of Numerical Analysis,II*, pages 16–351, Amsterdam, 1991. North-Holland, Amsterdam.

[Clé75]   P. Clément. Approximation by finite element functions using local regularization. *R.A.I.R.O. Anal. Numer.*, 9:77–84, 1975.

[DB05]   L. Demkowicz and A. Buffa. H^1, $H(\text{curl})$ and $H(\text{div})$-conforming projection-based interpolation in three dimensions. Quasi-optimal p-interpolation estimates. *Comput. Methods Appl. Mech. Engrg.*, 194(2-5):267–296, 2005.

[DBR01]   A. Düster, H. Bröker, and E. Rank. The p-version of the finite element method for three-dimensional curved thin walled structures. *Int. J. Num. Meth. Eng.*, 52:673–703, 2001.

[Dem07]   L. Demkowicz. *Computing with hp-adaptive finite elements. Vol. 1*. Chapman & Hall/CRC Applied Mathematics and Nonlinear Science Series. Chapman & Hall/CRC, Boca Raton, FL, 2007. One and two dimensional elliptic and Maxwell problems, With 1 CD-ROM (UNIX).

[DFY04]   M. Dauge, E. Faou, and Z. Yosibash. Plates and shells: Asymptotic expansions and hierarchical models. In R. de Borst E. Stein and R. Hughes T. J. editors, *Encyclopedia of Computational Mechanics*, volume I, pages 199–236. 2004.

[DKP+08]   L. Demkowicz, J. Kurtz, D. Pardo, M. Paszyński, W. Rachowicz, and A. Zdunek. *Computing with hp-adaptive finite elements. Vol. 2*. Chapman & Hall/CRC Applied Mathematics and Nonlinear Science Series. Chapman & Hall/CRC, Boca Raton, FL, 2008. Frontiers: three dimensional elliptic and Maxwell problems with applications.

[DL76]   G. Duvaut and J.-L. Lions. *Inequalities in Mathematics and Physics*. Springer-Verlag, Berlin Heidelberg New York, 1976.

[dV65]   B. Fraeijs de Veubeke. Displacement and equilibrium models in the finite element method. In O.C. Zienkiewicz and G. Holister, editors, *Stress Analysis*, pages 145–197. Wiley, New York, 1965.

[DW90]   M. Dryja and O. B. Widlund. Towards a unified theory of domain decomposition algorithms for elliptic problems. In *Third International Symposium on Domain Decomposition Methods for Partial Differential Equations (Houston, TX, 1989)*, pages 3–21. SIAM, Philadelphia, PA, 1990.

[FdV79] B. M. Fraeijs de Veubeke. *A course in elasticity*, volume 29 of *Applied Mathematical Sciences*. Springer-Verlag, New York, 1979. Translated from the French by F. A. Ficken.

[GOS03] M. Griebel, D. Oeltz, and M. A. Schweitzer. An algebraic multigrid method for linear elasticity. *SIAM J. Sci. Comput.*, 25(2):385–407 (electronic), 2003.

[GR86]   V. Girault and P. A. Raviart. *Finite Element Methods for Navier-Stokes Equations*. Springer-Verlag, Berlin Heidelberg New York Tokyo, 1986.

[Gri85] P. Grisvard. *Elliptic Problems in Non-Smooth Domains*. Pitman, Marshfield, Mass., 1985.

[Hac85] W. Hackbusch. *Multigrid methods and applications*, volume 4 of *Springer Series in Computational Mathematics*. Springer-Verlag, Berlin, 1985.

[Hip02] R. Hiptmair. Finite elements in computational electromagnetism. *Acta Numer.*, 11:237–339, 2002.

[HT00]   R. Hiptmair and A. Toselli. Overlapping and multilevel Schwarz methods for vector valued elliptic problems in three dimensions. In *Parallel solution of partial differential equations (Minneapolis, MN, 1997)*, volume 120 of *IMA Vol. Math. Appl.*, pages 181–208. Springer, New York, 2000.

[Joh87] C. Johnson. *Numerical solution of partial differential equations by the finite element method*. Cambridge University Press, Cambridge, 1987.

[KM87]   M. Kočvara and J. Mandel. A multigrid method for three-dimensional elasticity and algebraic convergence estimates. *Appl. Math. Comput.*, 23(2):121–135, 1987.

[Koi60] W. T. Koiter. A consistent first approximation in the general theory of thin elastic shells. In *Proc. Sympos. Thin Elastic Shells (Delft, 1959)*, pages 12–33. North-Holland, Amsterdam, 1960.

[Kra08]   J. K. Kraus. Algebraic multigrid based on computational molecules. II. Linear elasticity problems. *SIAM J. Sci. Comput.*, 30(1):505–524, 2007/08.

[KS99]   G. Karniadakis and S. Sherwin. *Spectral/hp Element Methods for CFD*. Oxford University Press, New York, Oxford, 1999.

[Man90]   J. Mandel. Iterative solvers by substructuring for the p-version finite element method. In *Spectral and high order methods for partial differential equations (Como, 1989)*, pages 117–128. North-Holland, Amsterdam, 1990.

[MH94]   J. E. Marsden and T. J. R. Hughes. *Mathematical foundations of elasticity*. Dover Publications Inc., New York, 1994. Corrected reprint of the 1983 original.

[Min51]   R. D. Mindlin. Influence of rotatory inertia and shear flexural motions of isotropic elastic plates. *J. Appl. Mech.*, 18:31–38, 1951.

[Mon03]   P. Monk. *Finite element methods for Maxwell's equations*. Claredon Press, Oxford, 2003.

[MW06]   K.-A. Mardal and R. Winther. An observation on Korn's inequality for nonconforming finite element methods. *Math.Comp.*, 75(253):1–6, 2006.

[Néd80]   J. C. Nédélec. Mixed finite elements in $\mathbb{R}^3$. *Numer. Math.*, 35:315–341, 1980.

[Néd86]   J. C. Nédélec. A new family of mixed finite elements in $\mathbb{R}^3$. *Numer. Math*, 50:57–81, 1986.

[Nit81]   J. A. Nitsche. On Korn's second inequality. *RAIRO Anal. Numér.*, 15(3):237–248, 1981.

[Pei91]   P. Peisker. A multigrid method for Reissner-Mindlin plates. *Numer. Math.*, 59(5):511–528, 1991.

[Pil08]   V. Pillwein. *Computer Algebra Tools for Special Functions in High Order Finite Element Methods*. PhD thesis, Johannes Kepler University Linz, Austria, 2008.

[PRS90]   P. Peisker, W. Rust, and E. Stein. Iterative solution methods for plate bending problems: multigrid and preconditioned cg algorithm. *SIAM J. Numer. Anal.*, 27(6):1450–1465, 1990.

[PW06]   J. E. Pasciak and Y. Wang. A multigrid preconditioner for the mixed formulation of linear plane elasticity. *SIAM J. Numer. Anal.*, 44(2):478–493 (electronic), 2006.

[PZ02]   J. E. Pasciak and J. Zhao. Overlapping Schwarz methods in $H(\operatorname{curl})$ on polyhedral domains. *J. Numer. Math.*, 10(3):221–234, 2002.

[QV97]   A. Quarteroni and A. Valli. *Numerical Approximation of Partial Differential Equations*. Springer-Verlag, Berlin, 1997.

[Rei45]   E. Reissner. The effect of transverse shear deformation on the bending of elastic plate models. *J. Appl. Mech.*, 12:69–76, 1945.

[Rei50]   E. Reissner. On a variational theorem in elasticity. *J. Math. Physics*, 29:90–95, 1950.

[RS85]   J. Ruge and K. Stüben. Efficient solution of finite difference and finite element equations by algebraic multigrid (AMG). In D. J. Paddon and H. Holstein, editors, *Multigrid Methods for Integral and Differential Equations*, IMA Conference Series, pages 169–212. Clarendon Press, Oxford, 1985.

[RS87]   J. W. Ruge and K. Stüben. Algebraic multigrid. In *Multigrid methods*, volume 3 of *Frontiers Appl. Math.*, pages 73–130. SIAM, Philadelphia, PA, 1987.

[RT77]   P.-A. Raviart and J. M. Thomas. A mixed finite element method for 2nd order elliptic problems. In *Mathematical aspects of finite element methods (Proc. Conf., Consiglio Naz. delle Ricerche (C.N.R), Rome, 1975)*, volume 606 of *Lecture notes in Mathematics*, pages 292–315, Berlin, 1977. Springer.

[Sch70]   H. A. Schwarz. Über einen Grenzübergang durch alternierendes Verfahren. *Vierteljahrsschrift der Naturforschenden Gesellschaft in Zürich*, 15:272–286, May 1870.

[Sch96]   C. Schwab. A-posteriori modeling error estimation for hierarchic plate models. *Numer. Math.*, 74:221–259, 1996.

[Sch97]   J. Schöberl. An advancing front 2D/3D mesh generator based on abstract rules. *Comput. Visual. Sci.*, 1:41–52, 1997.

[Sch98]   C. Schwab. *p- and hp-finite element methods*. Numerical Mathematics and Scientific Computation. The Clarendon Press Oxford University Press, New York, 1998. Theory and applications in solid and fluid mechanics.

[Sch99]   J. Schöberl. Multigrid methods for a parameter dependent problem in primal variables. *Numer. Math.*, 84(1):97–119, 1999.

[Sch01]   J. Schöberl. Commuting quasi-interpolation operators for mixed finite elements. Report isc-01-10-math, Texas A&M University, 2001.

[Sch08]   J. Schöberl. A posteriori error estimates for Maxwell equations. *Math. Comp.*, 77(262):633–649, 2008.

[SMPZ08] J. Schöberl, J. M. Melenk, C. Pechstein, and S. Zaglmayr. Additive Schwarz preconditioning for p-version triangular and tetrahedral finite elements. *IMA J. Numer. Anal.*, 28(1):1–24, 2008.

[SO97]   E. Stein and S. Ohnimus. Coupled model- and solution-adaptivity in the finite-element method. *Comput. Methods Appl. Mech. Engrg.*, 150(1-4):327–350, 1997. Symposium on Advances in Computational Mechanics, Vol. 2 (Austin, TX, 1997).

[SS06]   J. Schöberl and R. Stenberg. Multigrid methods for a stabilized Reissner-Mindlin plate formulation. Report a512, Helsinki University of Technology, 2006.

[Ste70]   E. M. Stein. *Singular integrals and differentiability properties of functions.* Princeton Mathematical Series, No. 30. Princeton University Press, Princeton, N.J., 1970.

[Ste86]   R. Stenberg. On the construction of optimal mixed finite element methods for the linear elasticity problem. *Numer. Math.*, 42:447–462, 1986.

[Ste88]   R. Stenberg. A family of mixed finite elements for the elasticity problem. *Numer. Math. 53, 513-538*, 1988.

[SZ80]   L. R. Scott and S. Zhang. Finite element interpolation of nonsmooth functions satisfying boundary conditions. *Math. Comp*, 54(190):483–493, 1980.

[SZ05]   J. Schöberl and S. Zaglmayr. High order Nédélec elements with local complete sequence properties. *COMPEL*, 24(2):374–384, 2005.

[TW59]   S. P. Timoshenko and S. Woinowsky-Krieger. *Theory of plates and shells.* Engineering Societies Monographs, New York: McGraw-Hill, 1959, 2nd ed., 1959.

[TW05]   A. Toselli and O. Widlund. *Domain decomposition methods – Algorithms and theory.* Springer-Verlag, Berlin Heidelberg New York, 2005.

[VB81]   M. Vogelius and I. Babuška. On a dimensional reduction method. I. The optimal selection of basis functions. *Math. Comp.*, 37(155):31–46, 1981.

[Wer02]   D. Werner. *Funktionalanalysis.* Springer, Berlin, Heidelberg, New York, 2002.

[Wie00]   C. Wieners. Robust multigrid methods for nearly incompressible elasticity. *Computing*, 64(4):289–306, 2000. International GAMM-Workshop on Multigrid Methods (Bonn, 1998).

[Zag06]   S. Zaglmayr. *High Order Finite Element Methods for Electromagnetic Field Computation.* PhD thesis, Johannes Kepler University Linz, Austria, 2006.

[ZZ87]   O. C. Zienkiewicz and J. Z. Zhu. A simple error estimator and adaptive procedure for practical engineering analysis. *Internat. J. Numer. Methods*, 24(2):337–357, 1987.

Die VDM Verlagsservicegesellschaft sucht für wissenschaftliche Verlage abgeschlossene und herausragende

Dissertationen, Habilitationen, Diplomarbeiten, Master Theses, Magisterarbeiten usw.

für die kostenlose Publikation als Fachbuch.

Sie verfügen über eine Arbeit, die hohen inhaltlichen und formalen Ansprüchen genügt, und haben Interesse an einer honorarvergüteten Publikation?

Dann senden Sie bitte erste Informationen über sich und Ihre Arbeit per Email an *info@vdm-vsg.de*.

Sie erhalten kurzfristig unser Feedback!

VDM Verlagsservicegesellschaft mbH
Dudweiler Landstr. 99   Telefon +49 681 3720 174
D - 66123 Saarbrücken   Fax   +49 681 3720 1749
www.vdm-vsg.de

Die VDM Verlagsservicegesellschaft mbH vertritt

Printed by Books on Demand GmbH, Norderstedt / Germany